DETAILED INTERPRETATION OF CONTEMPORARY ARCHITECTURAL DESIGN

当代建筑设计详解

佳图文化 主编

中国林业出版社

图书在版编目（ＣＩＰ）数据

当代建筑设计详解 . 3 / 佳图文化主编 . -- 北京：中国林业出版社 , 2018.10

ISBN 978-7-5038-9793-1

Ⅰ . ①当… Ⅱ . ①佳… Ⅲ . ①建筑设计－作品集－中国－现代 Ⅳ . ① TU206

中国版本图书馆 CIP 数据核字 (2018) 第 239854 号

中国林业出版社
责任编辑：李 顺 薛瑞琦
出版咨询： （010）83143569

出 版：中国林业出版社（100009 北京西城区德内大街刘海胡同 7 号）
网 站：http://lycb.forestry.gov.cn/
印 刷：固安县京平诚乾印刷有限公司
发 行：中国林业出版社
电 话：（010）83143500
版 次：2019 年 9 月第 1 版
印 次：2019 年 9 月第 1 次
开 本：889mm×1194mm 1 ／ 16
印 张：20
字 数：200 千字
定 价：298.00 元

Preface 前言

"Modern Chinese Architecture" series is a set of professional books that introduce and present modern Chinese architecture comprehensibly. It selects the excellent works with different design concepts and of different types that could reflect the level and development trend of contemporary Chinese architecture the most to represent the style and features. It revolves around the architectural works and analyzes further into overall planning, design concept, challenge, architectural layout, detailed design, relationship between the old and new in reconstruction, relationship between the building and the city, etc. thus to interpret the design essence from various angles. The selected works include office buildings, education buildings, commercial buildings, cultural buildings, hotel buildings, sports buildings, transportation buildings, healthcare buildings and so on. All the works are introduced with plans, elevations, sections, construction drawings, joint details and photos, presenting diverse architectural forms and details in a comprehensive and visual way for dear readers.

《当代中国建筑》系列图书为一套全面介绍和展示当代中国建筑的专业建筑类书籍。全书精选最能反映当代中国建筑水平和发展趋势的优秀作品，通过对这些不同类型、不同设计理念的建筑作品的展示和解读，反映当代中国建筑的风貌。全书内容围绕建筑设计案例展开，深入分析建筑案例的整体规划、设计构思、设计难点、建筑布局、细部设计、改扩建中新与老的联系、建筑与城市的关系等众多方面的问题，以期从各个角度展现建筑案例的设计精髓。所选案例囊括了办公建筑、教育建筑、商业建筑、文化建筑、酒店建筑、体育建筑、交通建筑、医疗建筑等众多建筑类型。案例中展示的平面图、立面图、剖面图、施工图、节点详图，以及建成实景图等，将不同的建筑形式以及建筑细节内容更为详尽、直观地呈现出来，以供关注当代中国建筑状况的读者借鉴、参考。

Contents 目录

Cultural Building 文化建筑

- 010 Rebuilt and Extension Project of Heibei Library
  河北省图书馆改扩建工程
- 020 Beichuan Earthquake Memorial Park Happiness Garden Exhibition
  北川抗震纪念园幸福园展览馆
- 026 New Anhui Museum
  安徽省博物馆新馆
- 034 Baoji Bronze Museum
  宝鸡市青铜器博物院
- 042 Li Yizhi Memorial Building and Park Planning
  李仪祉纪念馆建筑及园区规划设计
- 046 Pingdingshan Museum
  平顶山博物馆
- 054 Taicang Culture and Art Center
  太仓市文化艺术中心
- 066 Hangzhou Wushan Museum
  杭州吴山博物馆
- 072 China National Theatre and the Office
  中国国家话剧院剧场及办公楼

Cultural Building
文化建筑

Sports Building
体育建筑

078	Chongqing Science & Technology Museum 重庆科技馆
088	Xuzhou Music Hall 徐州音乐厅
094	Urban Planning Exhibition Hall, Liu'an, Anhui 安徽省六安市城市规划展览馆
098	Beijing International Flower Logistic Port 北京国际花卉物流港
104	Hebei Qian'an Cultural Convention and Exhibition Center 河北迁安市文化会展中心
110	Longyan Convention and Exhibition Center 龙岩市会展中心
116	Shenxianshu Community Service Center 神仙树社区服务中心
122	Water Park Reconstruction Project 水上公园改造工程
132	Tianjin Meijiang Convention Center 天津梅江会展中心
138	Xinjiang Science and Technology Museum Reconstruction and Extension Project 新疆科技馆改扩建工程

Contents 目录

- 142 Reconstruction and Extension Project of Guangdong Provincial Sun Yat-Sen Library
  广东省立中山图书馆改扩建工程
- 150 Ninghai Library
  宁海县图书馆
- 156 Hangzhou Cuisines Museum
  杭帮菜博物馆
- 160 Liuzhou Wonder Stone Museum
  柳州奇石馆

Sports Building 体育建筑

- 174 2010 Guangzhou Asian Games Swimming and Diving Hall of Provincial Venues
  2010 年广州亚运会省属场馆游泳跳水馆
- 182 South China Technology University Stadium, Guangzhou University City
  广州大学城华南理工大学体育馆
- 188 National Tennis Hall
  国家网球馆
- 196 Luoyang New District Sports Center Stadium
  洛阳新区体育中心体育场

202	Nansha Gymnasium 南沙体育馆
208	Putian Sports Center 莆田市体育中心
216	Shanxi Sports Center 山西省体育中心
224	Shenzhen University Games Center Stadium 深圳大运中心体育场
230	Shenzhen Bao'an Stadium 深圳市宝安体育场
238	Fujian Provincial Physical Rehabilitation and Employment Training Center for the Disabled 福建省残疾人体育康复就业培训中心
242	Dalian Stadium 大连市体育场
246	Fujian Physical Polytechnic College Track and Field Training Center 福建省体育职业技术学院田径训练馆
252	Liaoning Sports Training Center (Baiye Base) 辽宁省体育训练中心（柏叶基地）

Cultural Building
文化建筑

| Unique Appearance |
| 外观独特 |

| Local Characteristics |
| 地方特色 |

| Cultural Atmosphere |
| 人文精神 |

| Sustainability |
| 可持续性 |

KEY WORDS 关键词	Mosaic outside Decorations 马赛克外装
	Lattice Wall 花格墙
	Metal Louvres 金属百叶

Rebuilt and Extension Project of Heibei Library

河北省图书馆改扩建工程

FEATURES 项目亮点

The designer creates rustic and overwhelming local architectural characteristics in Hebei through ambitious and structured scale; the unique styled library highlights the solemn cultural atmosphere.

通过宏大、规整的比例尺度，展示图书馆的独特气质，彰显质朴、大气的河北本土建筑性格，凸显图书馆的庄重和文化氛围。

Location: Shijiazhuang, Hebei, China
Architectural Design: Hebei Architectural Design Institute Co., Ltd

Overview

The project with north-south axis in the center of cultural region as symmetrcal axis, the building extends to eastern and western side symmetrically, and the principle is the overall volume is not higher than reconstructed provincial museum.

项目地点：中国河北省石家庄市
设计单位：河北建筑设计研究院有限责任公司

项目概况

本工程整体形象以所在文化中心区域的南北中轴线为对称轴，向东西两侧基本对称展开，且总体体量以不高于北侧改扩建后的省博物馆总体体量为原则。

Site Plan 总平面图

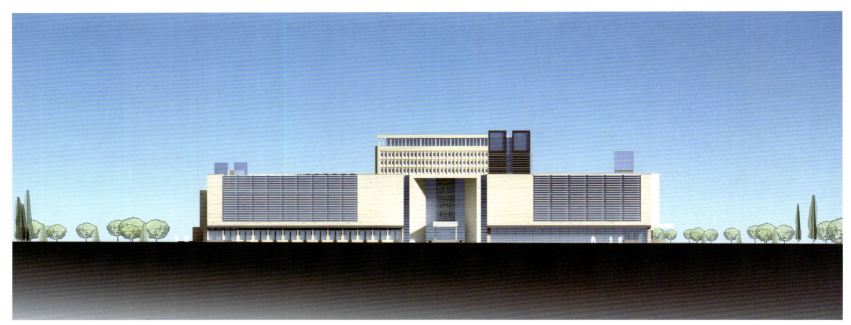

South Elevation 南立面图

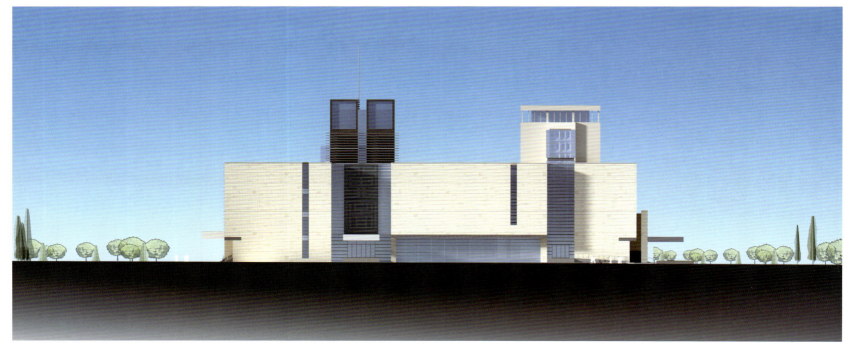

East Elevation 东立面图

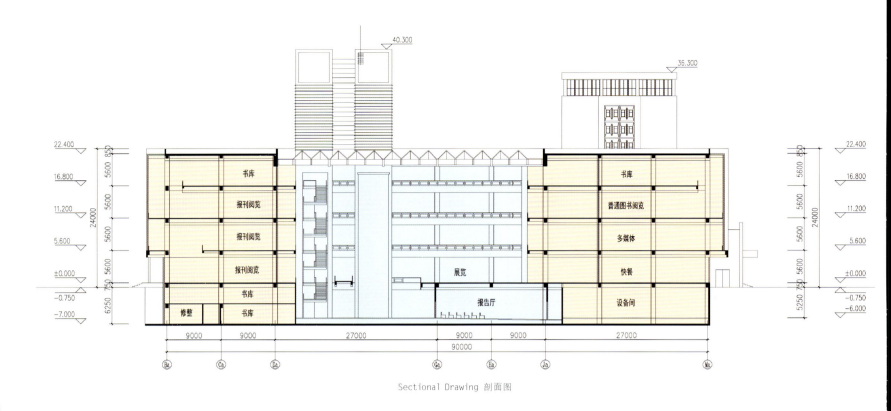

Sectional Drawing 剖面图

First Floor Plan 一层平面图

Second Floor Plan 二层平面图

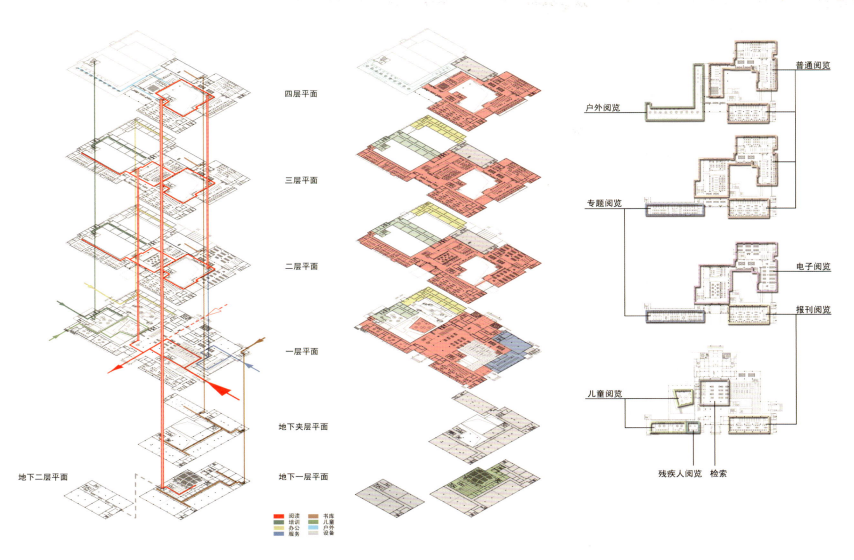

Architectural Design

The designer creates rustic and overwhelming local architectural characteristics in Hebei through ambitious and structured scale, the unique styled library highlights the solemn cultural atmosphere. The building integrates with surrounding environment; green spaces and public leisure places are added to provide citizens a large-scale cultural and entertainment place; to a certain extent, the urban environment has been improved and the urban quality has been promoted.

On shaping process, the designer tries try to show the characteristics of library and the distinct characteristics of times. In the aspect of building volume, the podium between new and old buildings achieves consistent scale relationship. According to outside decoration effects of new buildings, install metal louvers and glass for the old building in the form of "skin", hence old and new buildings achieve consistency in symmetric form.

Stack-room building has been remained, keeping its original steel-gray Mosaic outside decorations, and installing drawing metal mesh outside, which makes it possesses the fashionable expression of the existing buildings, but also fully demonstrates its original style. The sections of old buildings and the lattice wall on old buildings on the western side have been remained. The symmetrical "modern" lattice wall beside the old building develops into dramatic architectural expression. Meanwhile, besides full consideration of functional configuration and demand of building volume, the consistency of architectural style, image, color and other aspects between the extended library in the northern side and technology building in the southern side is also considered fully in the design of exterior facade shape, color, symmetry, balance and architectural style.

建筑设计

项目中，通过宏大、规整的比例尺度，展示图书馆的独特气质，彰显质朴、大气的河北本土建筑性格，凸显图书馆的庄重和文化氛围。建筑完美地融入周边环境，增加绿地和群众休闲活动的场所，为市民提供一个大型文化休闲场所，在一定程度上改善了城市环境并提升了城市品位。

在造型处理上，力图表现出图书馆建筑的性格特征和鲜明的时代特色。从体量上，新旧建筑之间在裙房部分取得一致的尺度关系。根据新建部分的外装饰效果，以"表皮"的形式，在旧建筑外安装金属百叶、玻璃，新旧建筑取得对称形式上的一致。

保留书库楼部分，保持其原有青灰色马赛克外装，在其外部安装拉延金属网，使其具有现阶段建筑应有的时尚表情的同时，又可全面展示其原有的风貌。保留旧建筑片段，保留西侧原建筑花格墙，并在其旁边新建与其对称的"现代"花格墙，形成颇具戏剧化的建筑表情。同时，除充分考虑本工程自身功能造型、体量需求外，在外立面形体、色彩、对称与均衡性及建筑风格方面，还充分考虑了与北侧改扩建后的省博物馆以及南侧的科技大厦楼群在整体建筑风格、形象、色彩等各层次的协调一致性。

KEY WORDS 关键词

- Landscape Platform 地景平台
- Unique Outlook 造型独特
- Cultural Connotation 文化内涵

Beichuan Earthquake Memorial Park Happiness Garden Exhibition
北川抗震纪念园幸福园展览馆

FEATURES 项目亮点

The design of rich sculpture "Whitehead" forms and echoes the theme of the earthquake memorial. The main building forms landscape platform by extending the Happiness Garden; simple and pure in its outlook, the main hall abstracts as "Whitehead" imagery.

建筑以富有雕塑感的"白石"造型设计呼应并共同形成抗震纪念园的主题。建筑主体以幸福园广场的延伸，形成地景平台；主展厅体量简洁、纯净，抽象为"白石"的意象。

Location: Aba Tibetan and Qiang Autonomous Prefecture, Sichuan, China
Architectural Design: Architectural Design and Research Institute of Tsinghua University, Ltd.
Land Area: approximately 2,000 m²
Floor Area: 2,221.7 m²
Plot Ratio: 0.17

项目地点：中国四川省阿坝藏族羌族自治州
设计单位：清华大学建筑设计研究院有限公司
用地面积：约 2 000 m²
建筑面积：2 221.7 m²
容积率：0.17

Overview

The exhibition is a theme building of the Happiness Garden, an important carrier to display items, record and spread the relief of Beichuan earthquake as well as the optimistic spirit of reconstruction the homeland. It also serves as a place for public communication and leisure. The exhibition has two layers with a height of 13.6 m.

项目概况

展览馆是幸福园的主题建筑，是陈列展览、纪录传播北川人民抗震救灾、重建家园乐观精神的重要载体，同时也是市民交流、沟通、休闲的场所。展览馆为两层，高度为 13.6 m。

Architectural Design

The design of rich sculpture "Whitehead" forms and echoes the theme of the earthquake memorial. The main building forms landscape platform by extending the Happiness Garden; simple and pure in its outlook, the main hall abstracts as "Whitehead" imagery. By using modern techniques subtly expressing of the traditional Qiang culture, the design metaphor "sacred", "asylum", "good luck" pray for the new Beichuan; the combination of open platform and square green landscape provides a cordial and harmonious urban public living space, expressing hope for a happy life in the future.

KEY WORDS 关键词

- **Square Volume** 方形体量
- **Unique Appearance** 独特外观
- **Huizhou Villages** 徽州村落

New Anhui Museum
安徽省博物馆新馆

FEATURES 项目亮点

With the concept "four waters go together", the core parts of the building form a public hall opening to the sky and a three-dimensional Huizhou village through turning, collusion, overlapping wall slices to define the space.

建筑核心部位取意"四水归堂",形成向天开敞的公共大厅,通过一系列转折勾连、错动交叠的片墙限定空间,形成一个立体的徽州村落。

Location: Hefei, Anhui, China
Architectural Design: Architectural Design and Research Institute, South China University of Technology
Land Area: 28,100 m²
Total Floor Area: 41,430 m²
Building Height: 37.7 m
Plot Ratio: 1.47

项目地点:中国安徽省合肥市
设计单位:华南理工大学建筑设计研究院
用地面积:28 100 m²
总建筑面积:41 430 m²
建筑高度:37.7 m
容积率:1.47

Architectural Design

The square building body displays space constitution and forms a continuous volume. The three-dimensional architectural practices demonstrate the power which is hardly to display in large-scale venues in dense urban areas. Its organic unity of static square outline and dynamic 3D structure shows a distinct, unique iconic image both from a nearby view or from afar. The overall spatial layout is clearly divided into three spheres and the three-dimensional configuration generates a complexity and diversity public space systems.

With the concept "four waters go together", the core parts of the building form a public hall opening to the sky and a three-dimensional Huizhou village through turning, collusion, overlapping wall slices to define the space. Through guidance and dislocation the wall slices, the public hall forms a series of large, medium, small space and "roadway". The escalators arranged in a gripping piece wall and sky-guided "roadway" circle to reach the layers.

建筑设计

整体方形的建筑体量尽显空间的构成方式,形成可以展开的虚实相间的连续整体,一气呵成的立体构成手法在大尺度的新区场地中尽显城市密集区域的建筑难以展现的力量,静态方正轮廓与动态立体构成的有机统一,无论在车行远观还是步行近观的尺度下,都呈现出鲜明、独特的标志性形象。建筑整体空间布局清晰地分成三个圈层,而立体构成的手法又形成具有复杂性与多样性的公共空间系统。

建筑核心部位取意"四水归堂",形成向天开敞的公共大厅,通过一系列转折勾连、错动交叠的片墙限定空间,形成一个立体的徽州村落。片墙的引导与错动使公共大厅串联起一系列大、中、小空间与"巷道",自动扶梯布置在片墙夹持、天光引导的"巷道"中,环绕盘旋到达各层空间。

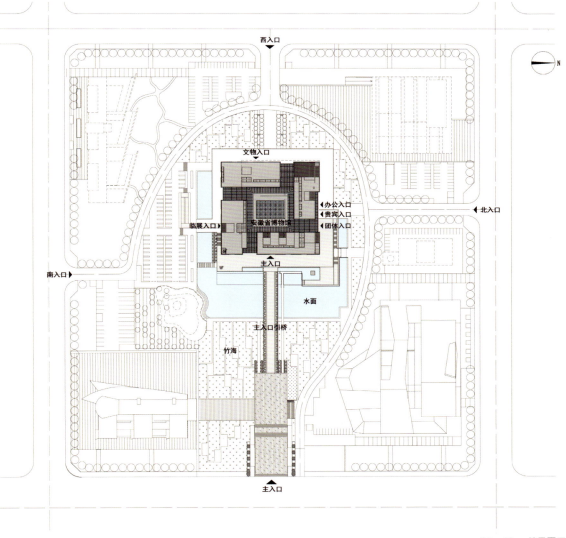

Site Plan 总平面图

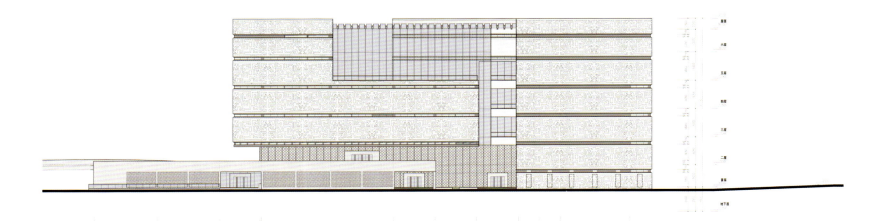

Elevation 1 立面图 1

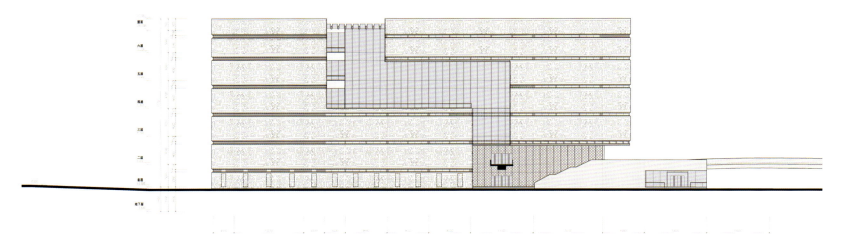

Elevation 2 立面图 2

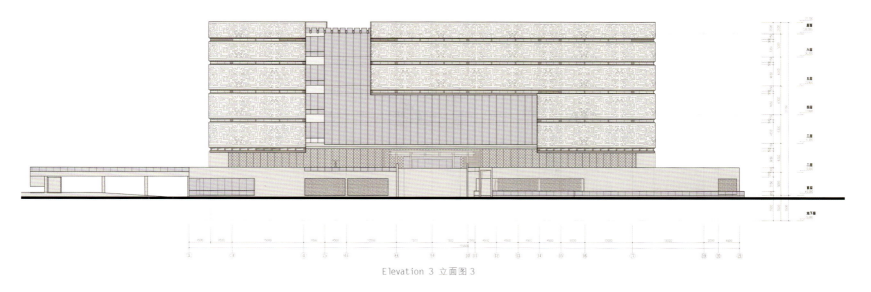
Elevation 3 立面图 3

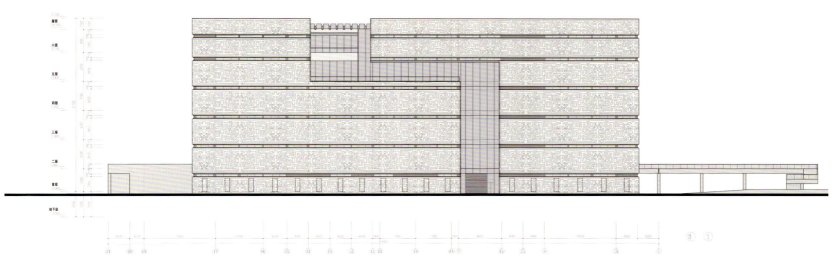
Elevation 4 立面图 4

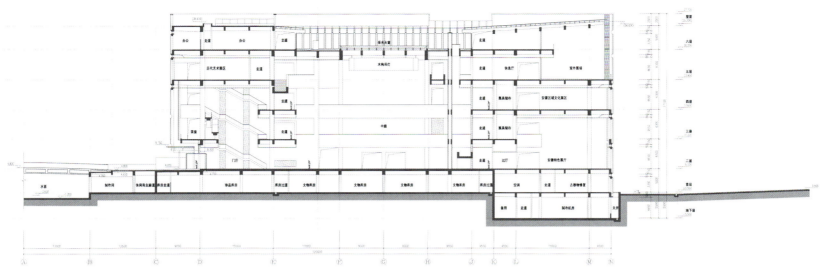

Section 1-1 1-1 剖面图

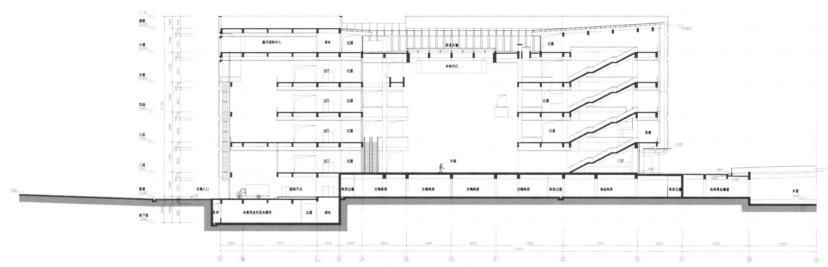

Section 2-2 2-2 剖面图

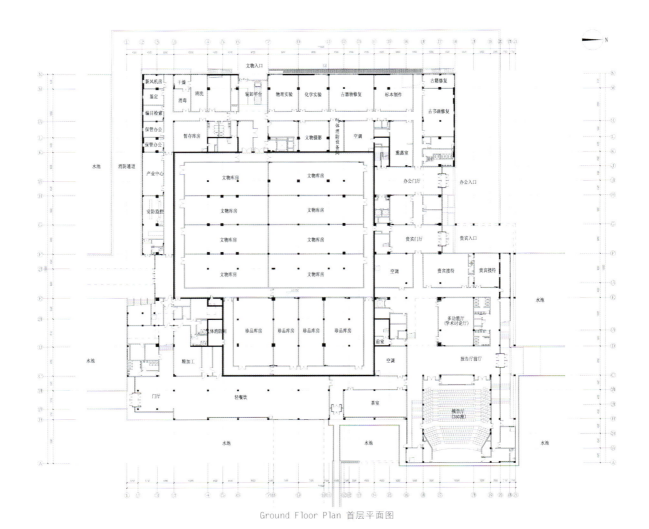

Ground Floor Plan 首层平面图

东南面外观夜景

KEY WORDS 关键词	Imposing Vigour 气势雄伟
	Unique Appearance 造型独特
	Chic Style 新颖别致

Baoji Bronze Museum
宝鸡市青铜器博物院

FEATURES 项目亮点

Baoji Bronze Museum main building is designed in the shape of the bronze mirror with a unique combination of Shek Kwu—Shek Kwu as the base, bronze mirror on the top, decorated with typical Western Zhou phoenix pattern; the design metaphor is to prominent Zhou and Qin wind, stone rhyme.

宝鸡青铜器博物院主楼的设计，在造型上把石鼓与铜镜巧妙结合——以石鼓为基座，以铜镜为顶面，饰以典型的西周凤鸟纹，旨在突出周秦之风、金石之韵。

Location: Baoji, Shaanxi, China
Architectural Design: Architectural Design and Research Institute of Tianjin University
Building Scale: 34,788 m²
Total Land Area: 34,788 m²
Greenland Area: 251,891 m²
Plot Ratio: 0.69
Green Coverage Ratio: 52%

项目地点：中国陕西省宝鸡市
设计单位：天津大学建筑设计研究院
建筑规模：34 788 m²
总用地面积：34 788 m²
绿地面积：251 891 m²
容积率：0.69
绿地率：52%

Overview

Baoji Bronze Museum is located in China Shek Kwu Park. The main building is a unique style of "platform five tripod" shape, imposing, novel, concentrated profound meaning of Western Zhou Ding system. The project is listed as a milestone in the history of Chinese bronze collections, a landmark in the western city of Baoji.

项目概况

宝鸡市青铜器博物院位于中华石鼓园内。主体建筑为风格独特的"平台五鼎"造型，气势雄伟，新颖别致，浓缩了西周列鼎制度的深刻内涵。项目被列为中国青铜器收藏史上的一个里程碑，是西部重镇宝鸡的标志性建筑。

Site Plan 总平面图

Elevation 立面图

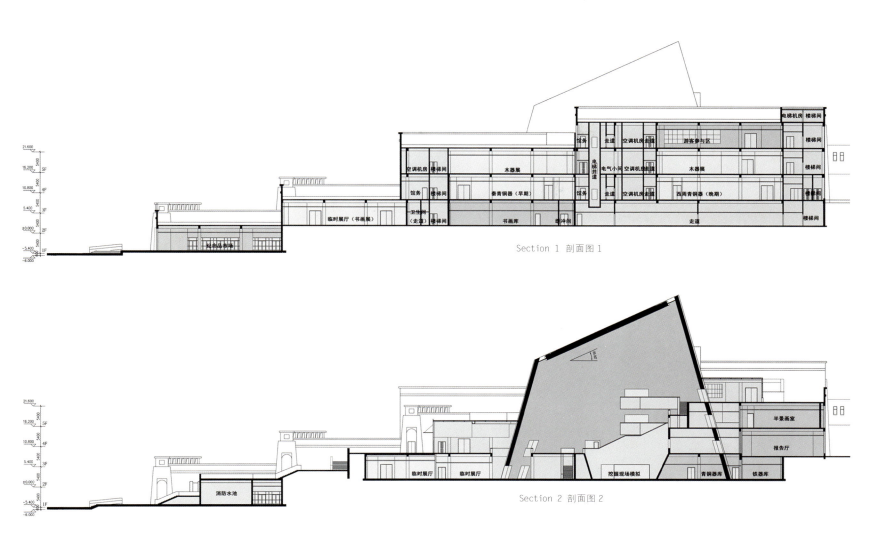

Section 1 剖面图 1

Section 2 剖面图 2

Architectural Design

Baoji Bronze Museum main building is designed in the shape of the bronze mirror with a unique combination of Shek Kwu—Shek Kwu as the base, bronze mirror on the top, decorated with typical Western Zhou phoenix pattern; the design metaphor is to prominent Zhou and Qin wind, stone rhyme.

Baoji Bronze Museum is divided into five main buildings, using a high-step door, bronze and dirt architectural language; the building metaphor is the history and culture of Baoji's respected position in the ancient Chinese civilization and also a perfect combination of Shek Kwu culture and Bronze culture.

建筑设计

宝鸡青铜器博物院主楼的设计，在造型上把石鼓与铜镜巧妙结合——以石鼓为基座，以铜镜为顶面，饰以典型的西周凤鸟纹，旨在突出周秦之风、金石之韵。

宝鸡青铜器博物院主体建筑分为五层，建筑形象运用了高台门阙、青铜后土的建筑语言，寓意着宝鸡悠久的历史文化在中国古代文明中的尊崇地位，同时也完美地结合了石鼓文化与青铜文化。

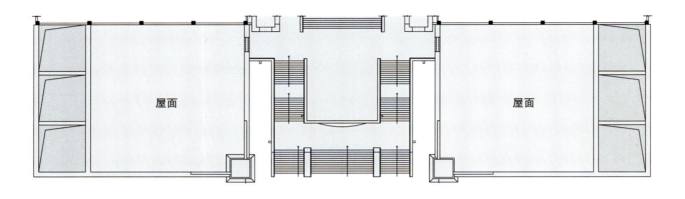

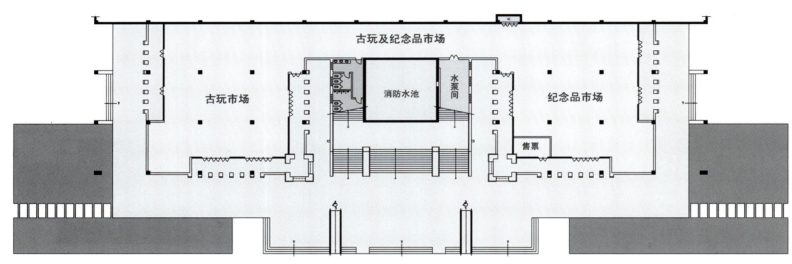

Ground Floor Plan 首层平面图

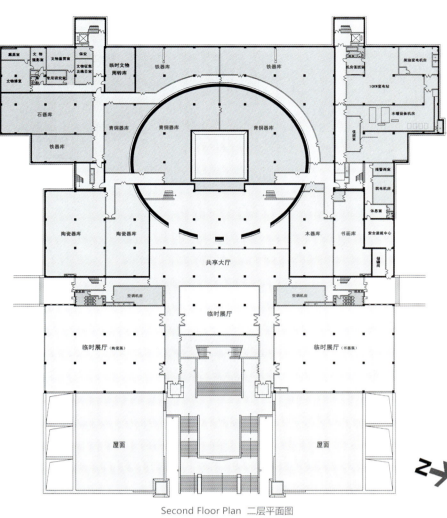

Second Floor Plan 二层平面图

KEY WORDS 关键词

Unique Modeling 造型独特
Reasonable Layout 布局合理
Main Commemorative 主体纪念

Li Yizhi Memorial Building and Park Planning
李仪祉纪念馆建筑及园区规划设计

FEATURES 项目亮点

By setting the memorial building as its center, the park planning exhibits water conservation and Li Yizhi's life stories. The planning emphasizes on memorial park atmosphere.

整个园区的规划设计以纪念馆建筑为中心，以展览水利文化和李仪址生平等为重心，强调纪念性的园区氛围。

Location: Xianyang, Shaanxi, China
Architectural Design: Northwest China Architecture Design & Research Co., Ltd.
Total Floor Area: 3,000 m²
Building Coverage Ratio: 0.051
Plot Ratio: 0.056

项目地点：中国陕西省咸阳市
设计单位：中国建筑西北设计研究院有限公司
总建筑面积：3 000 m²
建筑覆盖率：0.051
容积率：0.056

Overview

Located in Wangqiao Town, Jingyang County in Shaanxi Province, Li Yizhi Memorial took a year and a half to build, it is Shaanxi Province's first comprehensive theme memorial combining the history and culture of water conservancy development and water science popularization.

项目概况

李仪祉纪念馆位于陕西省泾阳县王桥镇，耗时一年半修建，是陕西省首座集水利发展史、水利文化展示、水利科学普及于一体的综合性主题纪念馆。

Design Principles

By setting the memorial building as its center, the park planning exhibits water conservation and Li Yizhi's life stories. The planning emphasizes on memorial park atmosphere. As the core of the park, the memorial building gives more attention on exhibition and memorial function; at the same time, it shows respect to Li Yizhi. It carries out the heavy responsibility on water conservation culture and is the most significant "landscape" and landmarks.

设计原则

整个园区的规划设计以纪念馆建筑为中心，以展览水利文化和李仪址生平等为重心，强调纪念性的园区氛围，纪念馆在整个园区中居核心地位，在表达对李仪址陵园尊重的同时更多地赋予其展览和纪念的功能，同时肩负传播水利文化的重任。建筑本身也是园区中最显著的"景观"和地标。

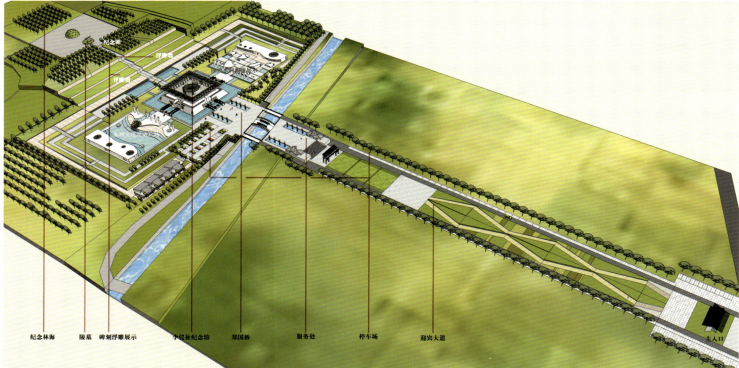

纪念林海　陵墓　碑刻浮雕展示　小说社纪念馆　郑国桥　服务处　停车场　迎宾大道　　　　主入口

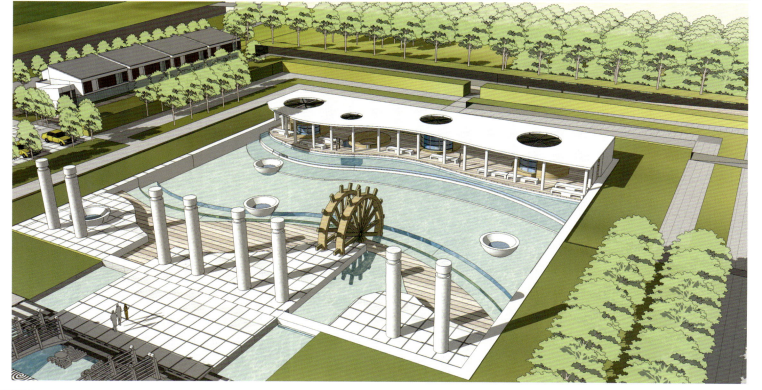

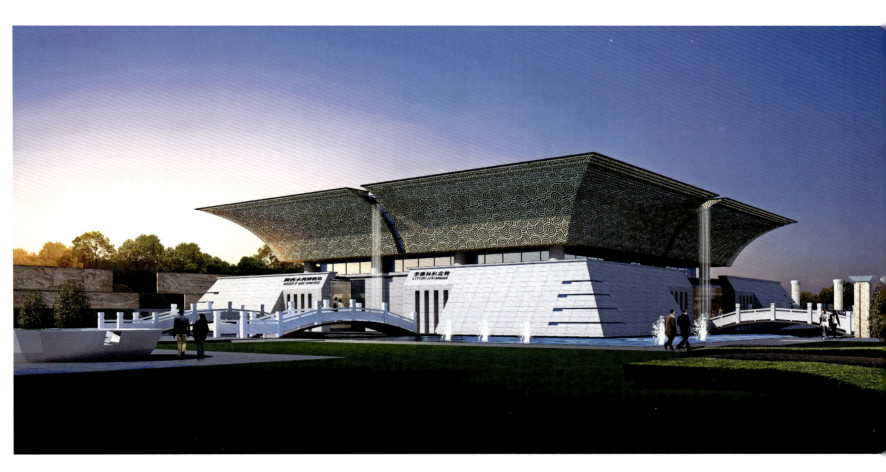

KEY WORDS 关键词

| Yellow Facade 黄色立面 |
| Exotic Appearance 外观奇异 |
| Reasonable Layout 布局合理 |

Pingdingshan Museum
平顶山博物馆

FEATURES 项目亮点

The facade develops into densely arranged closed surface, and closely connects the function of indoor exhibition space; the interspaces left by closed surface decorate the light, are like the stars in history, which are plain and full of vitality.

立面形成密集排列的封闭界面，与室内展区的功能紧密联系，而这些封闭的界面中留有的些许空隙，点缀星星点点之灯片，如同历史长河中的繁星，朴素而又充满活力。

Location: Pingdingshan, Henan, China
Architectural Design: Architectural Design & Research Institute of Tsinghua University Co., Ltd
Land Area: 320,000 m²
Total Floor Area: 30,074 m²

项目地点：中国河南省平顶山市
设计单位：清华大学建筑设计研究院有限公司
用地面积：320 000 m²
总建筑面积：30 074 m²

Planning and Layout

The function regionalization is clear and concise, east as the main pedestrian entrance, public space, transport space and service space are organized through combining entrance and entrance lobby; the atrium exhibition hall is on the north and the south side, and is divided into 3 parts: exhibition on the first floor, historical relics showroom on the second floor, Chinese folk, natural science and technology and urban planning showroom on the third floor; the ceiling height is 7 m, offering great convenience for exhibition. Offices and service rooms are arranged on west side.

规划布局

博物馆功能区划简洁、明了，东侧作为主要的人流入口，结合入口位于中部的入口大堂组织公共空间、交通空间、服务空间；南北两侧中庭展厅分成三部分：一层是临时展厅，二层是历史文物陈列，三层分别为中国民俗陈列、自然科技陈列及城市规划陈列，净高 7 m 的空间为展览提供巨大便利。建筑西侧为后勤办公及服务用房。

Facade Design

The facade develops into densely arranged closed surface, and closely connects the function of indoor exhibition space; the interspaces left by closed surface decorate the light, are like the stars in history, plain and full of vitality.

立面设计

立面形成密集排列的封闭界面，与室内展区的功能紧密联系，而这些封闭的界面中留有的些许空隙，点缀星星点点之灯片，如同历史长河中的繁星，朴素而又充满活力。

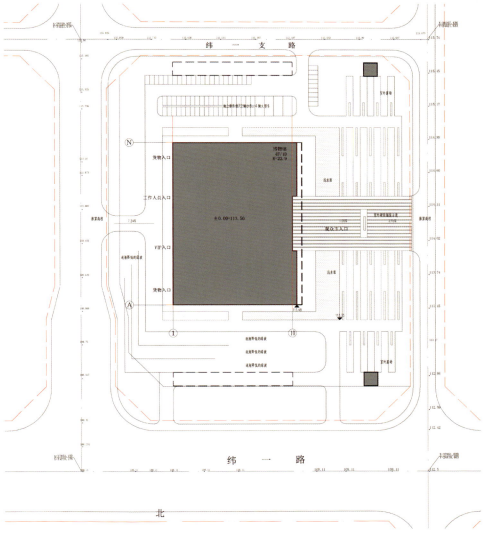

Site Plan 总平面图

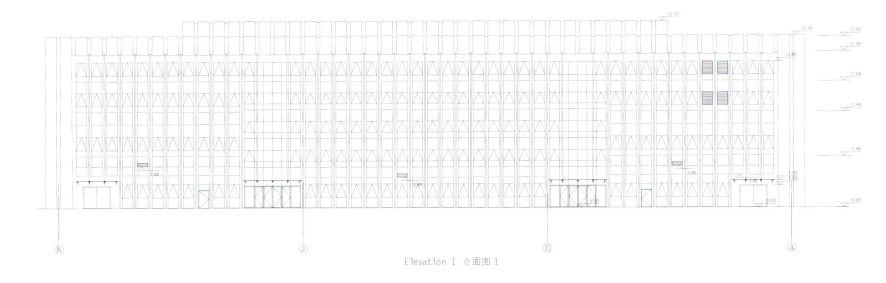

Elevation 1 立面图 1

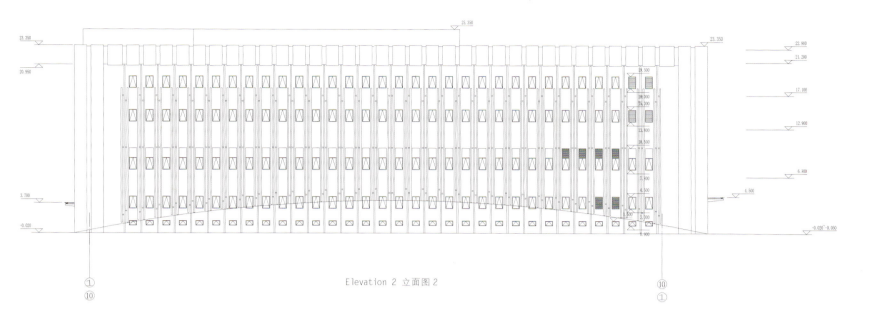

Elevation 2 立面图2

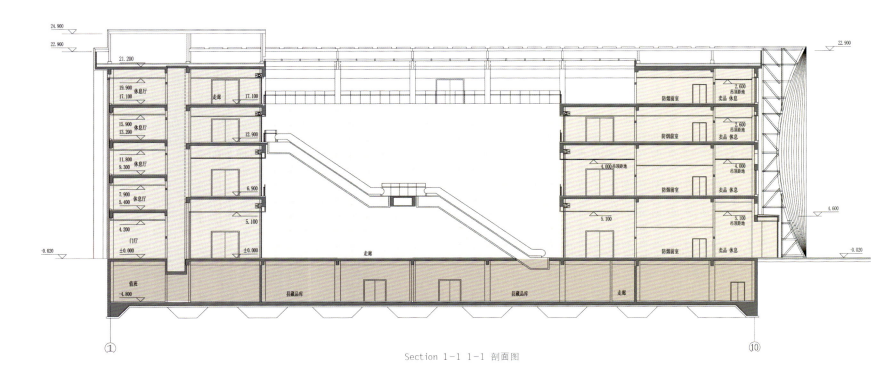

Section 1-1 1-1 剖面图

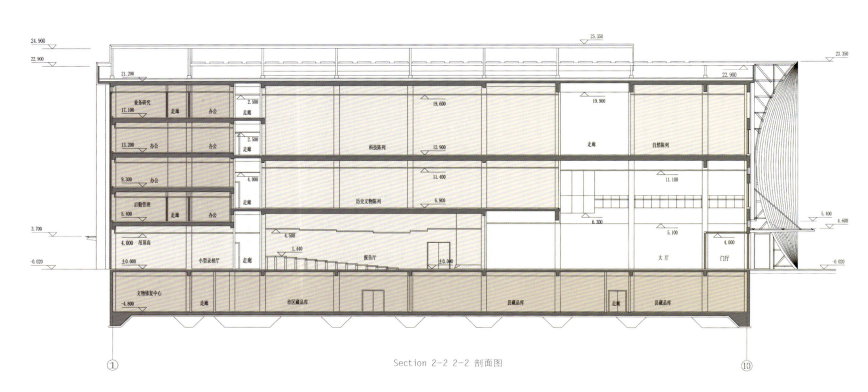

Section 2-2 2-2 剖面图

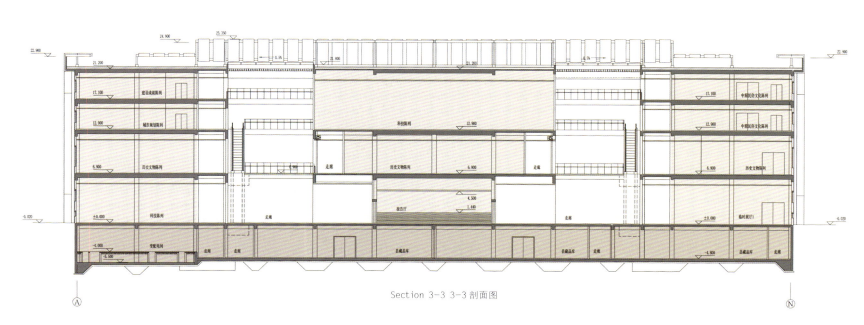

Section 3-3 3-3 剖面图

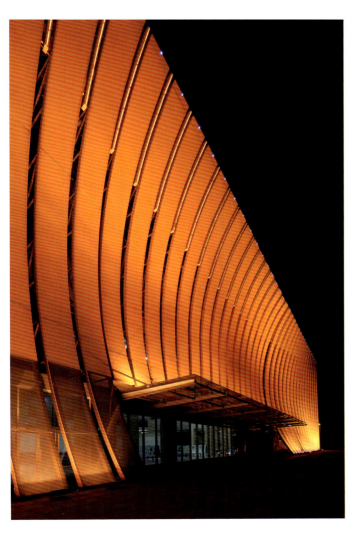

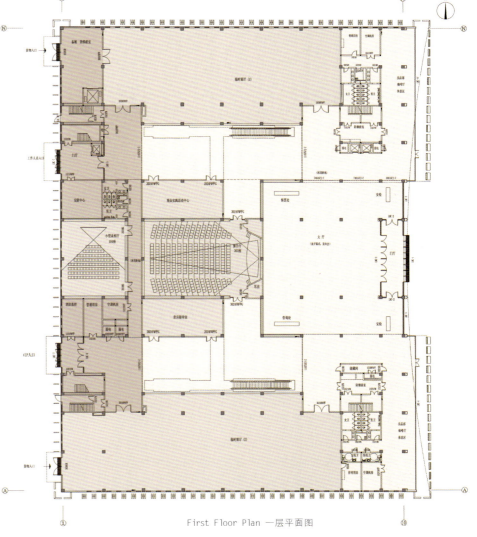

First Floor Plan 一层平面图

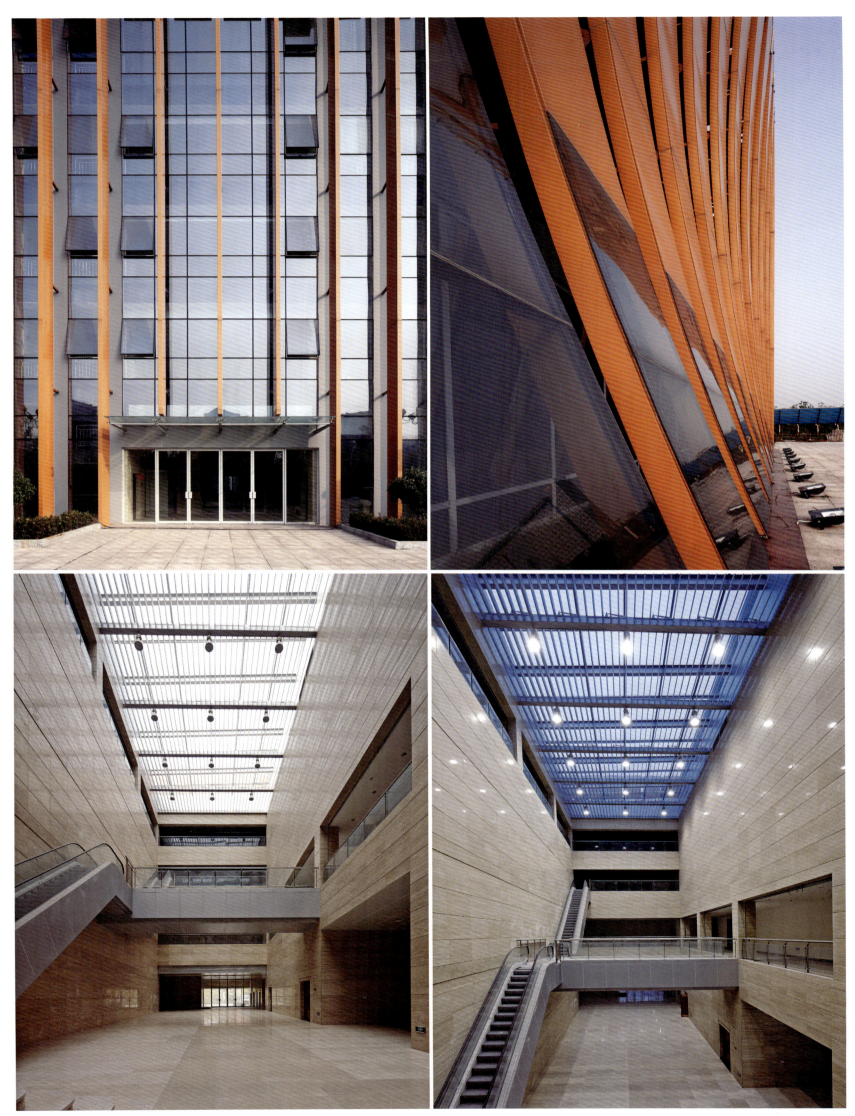

KEY WORDS 关键词

- Glass Façade 玻璃表皮
- Cable-membrane Curtain Wall 索网幕墙
- Green Roof 绿化屋面

Taicang Culture and Art Center
太仓市文化艺术中心

FEATURES 项目亮点

The design of this culture and art center aims to emphasize fusion of architecture and regional environment, tries to give humanity and cultural characteristics to the building, and embodies that the cultural building is consistent in traditional cultural and spiritual level.

设计强调建筑与区域环境的融合，极力赋予建筑人性和文化特征，体现文化建筑与传统文化在精神层面的一致。

Location: Taicang, Jiangsu, China
Land Area: 13,500 m²
Floor Area: 25,990 m²

项目地点：中国江苏省太仓市
用地面积：13 500 m²
建筑面积：25 990 m²

Overview

Taicang Culture and Art Center consists of grand theatre and cultural center, with the total site area of 13,500 m². The entire site is long in north and south, and narrow in east and west. Due to the limitation of the site, the grand theatre and cultural center expand from north to south; the eastern side that faces civil culture square is the main entrance.

项目概况

太仓市文化艺术中心由大剧院和文化馆两部分组成，总用地面积13 500 m²。整个场地南北长，东西窄。针对用地的限制，设计将大剧院和文化馆呈南北并置展开，以面向市民文化广场的东向为一院一馆的主入口。

Design Concept

The image position of Taicang Culture and Art Center is not pursuit of greatness, novelty, shocking and variability in traditional cultural building design, but expresses a particular jiangnan culture through inner emotions, shows "small", "light", "thin" and "elegant" Jiangnan culture inadvertently.

The design of this culture and art center aims to emphasize fusion of architecture and regional environment, tries to give humanity and cultural characteristics to building, and embodies that the cultural building is consistent in traditional culture spiritual level. On the aspect of architectural techniques, designers pay attention to reflect the elegant atmosphere of modern architecture space, and show humanistic care through enriching details.

设计构思

太仓市文化艺术中心的形象定位不追求一般意义上文化建筑造型的宏大与新奇、震撼与变化，而是通过内心情感来表现一种特定的江南文化，似不经意间展现"小"、"轻"、"细"、"雅"的文化江南。

文化艺术中心的设计宗旨为强调建筑与区域环境的融合，极力赋予建筑人性和文化特征，体现文化建筑与传统文化在精神层面的一致。同时，在建筑手法上，注重营造现代建筑空间的典雅氛围，通过丰富的细节表现人文关怀。

Site Plan 总平面图

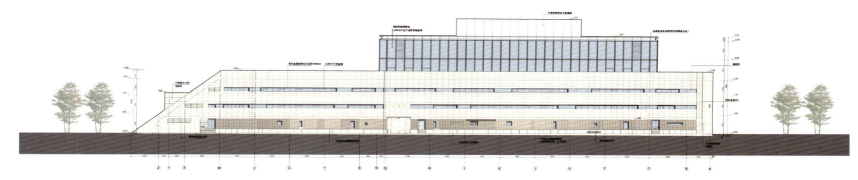
West Elevation 西立面图

South Elevation 南立面图

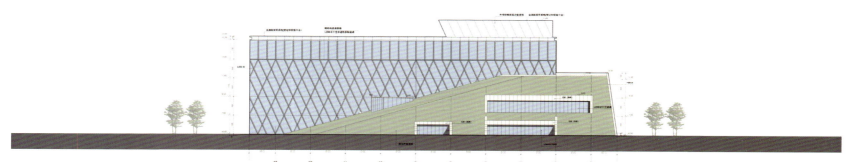
North Elevation 北立面图

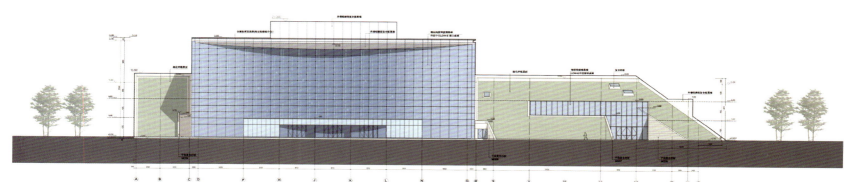
East Elevation 东立面图

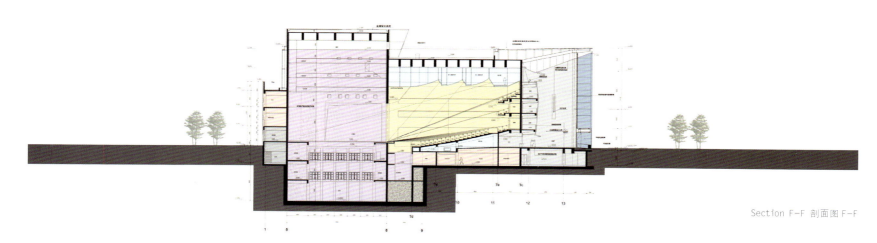
Section F-F 剖面图 F-F

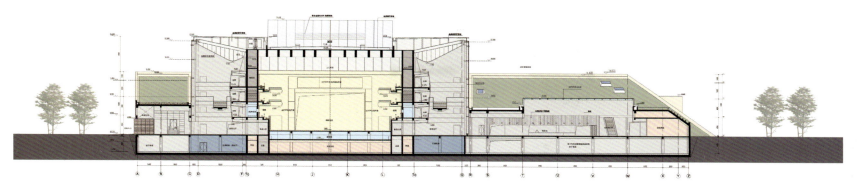

Section B1-B1 剖面图 B1-B1

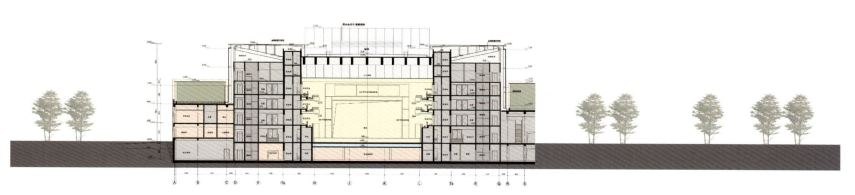

Section B2-B2 剖面图 B2-B2

057

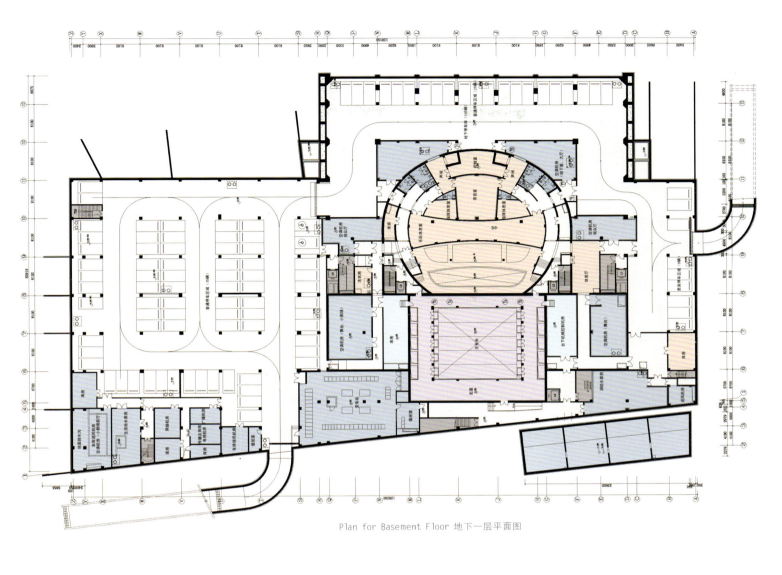

Plan for Basement Floor 地下一层平面图

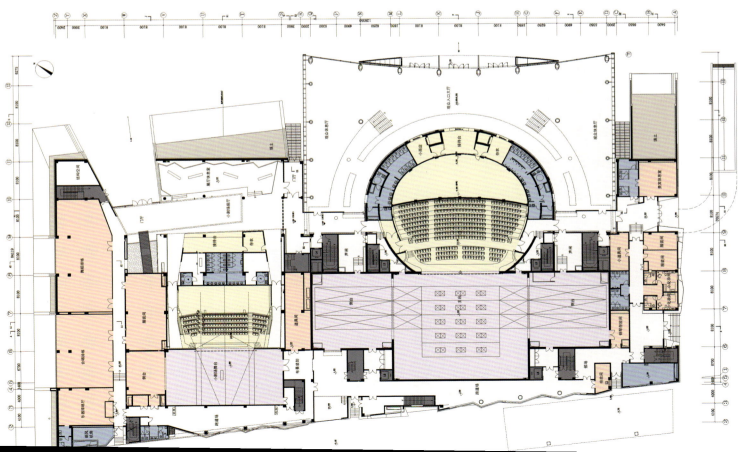

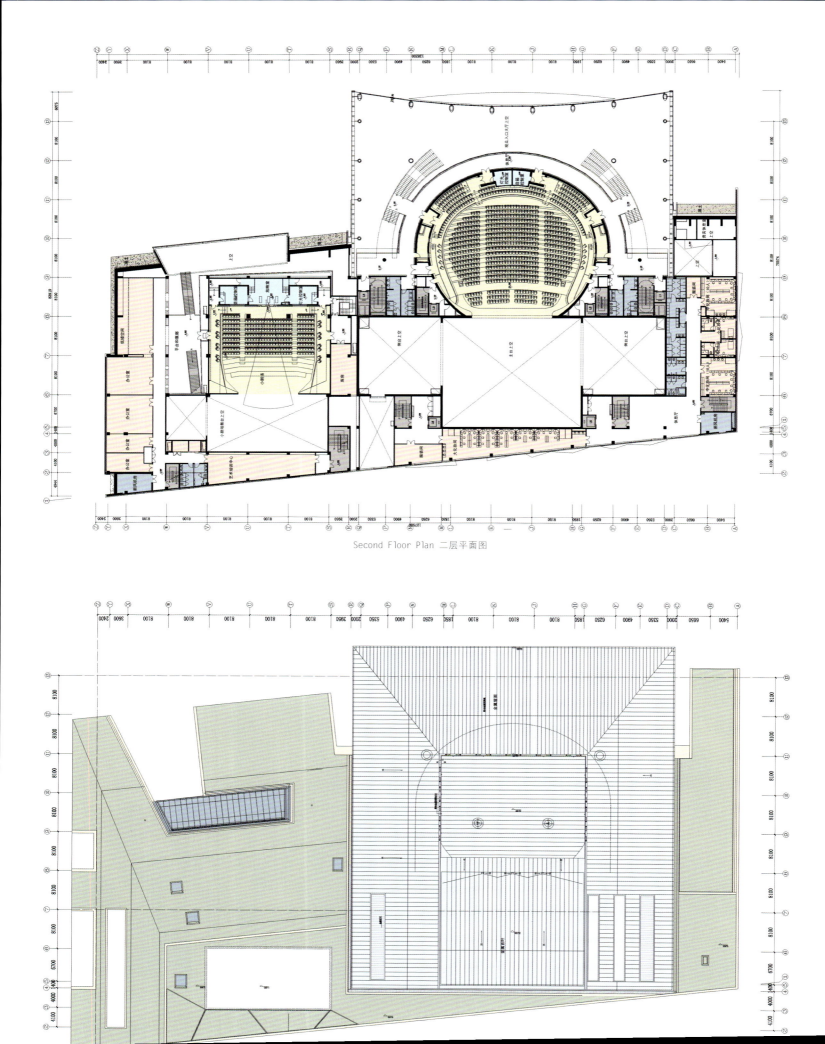

Second Floor Plan 二层平面图

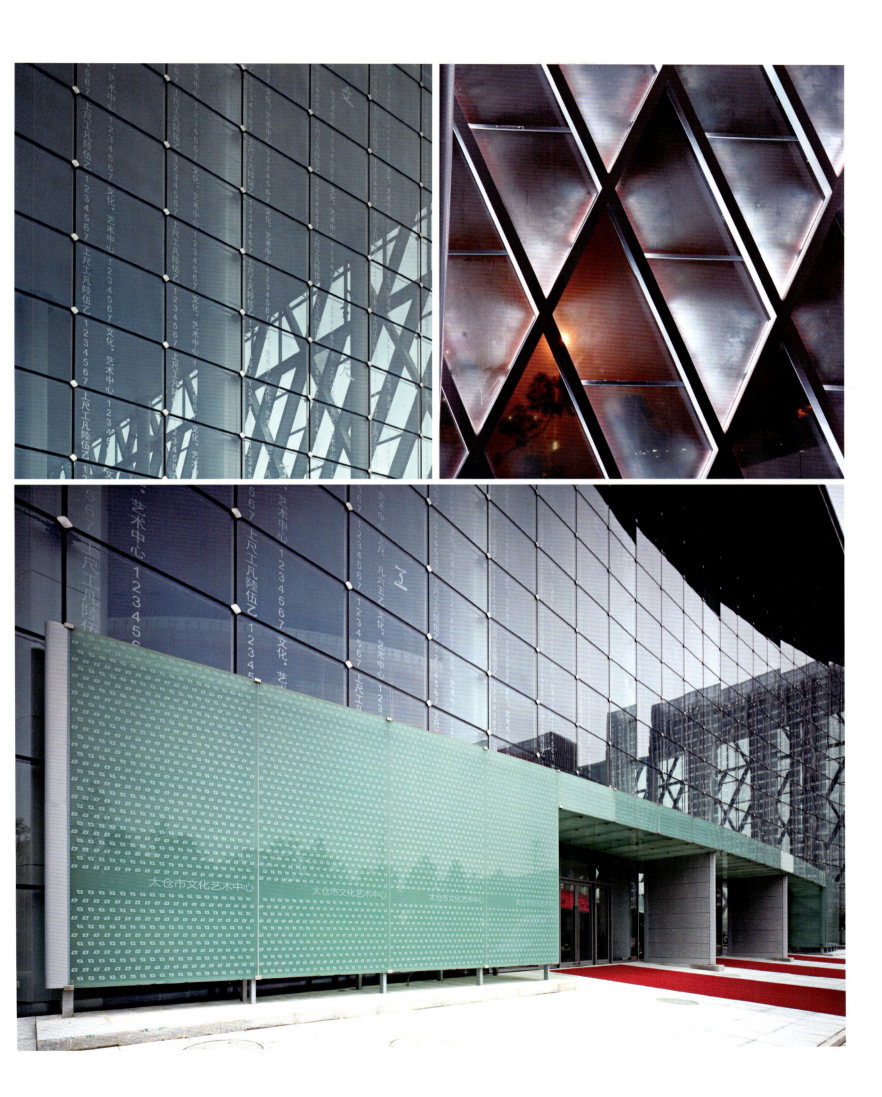

Architectural Design

Due to the limitation of the site, blanking technique is used in building design, which deliberately narrows building volume. Nearly full glass facade of main grand theater weakens the visual impact of building volume, while cultural center and other accessory hall hidden under the sloping green roof, only "small vitreous" of cultural center entrance hall leans out of grass, giving person space cueing. On the aspect of technique processing, rich varied architecture form with primary and auxiliary, with advance and retreat develops into form relaxed urban space.

The eastern facade as the main facade of building, the spreading cable-membrane curtain wall along the concave arc is similar to the curtain pulled open slowly; it is also like the music score with the jumping notes and melodies, giving extreme play to the role of light glass. Meanwhile, cable-membrane curtain borrows slope greening as the background, which makes architectural style more implicative, as if it integrates with the eastern city square gradually, therefore, dialogue on space achieved. As the southern facade and western facade facing urban avenue, the two boundaries of city are strengthened. The base of the grand theatre made up of vitreous and stone material emphasizes the continue relationship between modern culture and traditional culture. In the texture of glass curtain wall and stone curtain wall, parallelogram pattern as motif appears again and again, and meticulous and eclectic arrangement conveys a vibrant dynamic innovation.

建筑设计

由于用地的限制，建筑设计采用消隐的手法，刻意缩小建筑体量。主体大剧院几乎全玻璃的表皮弱化了建筑体量的视觉冲击；而文化馆与其他附属用房隐退于倾斜的绿化屋面之下，只有文化馆入口门厅的"小玻璃体"探出草地，给人以空间提示。在手法处理上体现建筑形体的一主一辅、一进一退的丰富变化，形成舒缓的城市空间。

文化中心以东立面作为建筑的主立面，沿内凹弧线轻盈展开的索网幕墙形似徐徐拉开的大幕，又仿佛注满跳跃音符与旋律的乐谱，将玻璃的轻盈发挥到极致。与此同时，索网幕墙以斜坡绿化为背景，使建筑风格更加轻柔、含蓄，仿佛渐渐地与东侧绿化市民广场融为一体，从而完成空间上的对话。南立面与西立面由于面朝城市道路，设计中刻意加强了这两个城市界面的延续。大剧院的玻璃体与石材形成的基座强调了现代文化与传统文化的承接关系。在玻璃幕墙和石材幕墙的肌理中，以平行四边形为母题的图案重复出现，细致而不拘一格的排列方式传达了一种充满活力动感的创新精神。

KEY WORDS 关键词	Natural Environment 自然环境
	Building Scale 建筑尺度
	Humanity Texture 人文肌理

Hangzhou Wushan Museum
杭州吴山博物馆

FEATURES
项目亮点

Principal axis of the building complies with landform, and develops into continuous and rhythmic spatial pattern. Additional courtyard and the terrace make the building and natural environment permeate and blend each other.

建筑主轴顺应山势逐步抬高，形成整体连续、张弛有度的空间格局。通过穿插其间的庭院和观景平台使自然环境和建筑完美融合。

Location: Hangzhou, Zhejiang
Land Area: 7,000 m²
Floor Area: 6,082 m²

项目地点：浙江省杭州市
用地面积：7 000 m²
建筑面积：6 082 m²

Overview

Wushan Museum is located in the southeastern side of Wushan square, where is the important connection of Tianfeng historical block and Qinghefang historical block in Wushan.

项目概况

吴山博物馆位于杭州吴山广场东南侧，是吴山天风、清河坊历史街区两大旅游景区的重要结合点。

Design Concept

Designers hope to restore the rupturing natural mountain through integration, and make the site of the museum stretching into a whole with natural texture. Meanwhile, they also hope that the building can coordinate with the natural disposition of Qinghefang block through scattered layout and appropriate scale, develop into natural continuation of original urban humanities, and create a unique architectural image that connects all the surroundings.

设计构思

设计师希望通过整合修复断裂的自然山体肌理，使博物馆基地与自然肌理连绵成一个整体；同时也希望通过分散的布局、适宜的尺度，很好地与清河坊街区的城市肌理相协调，形成原有城市人文肌理的自然延续，以此创造与吴山山麓相关联且富有特色的整体建筑形象。

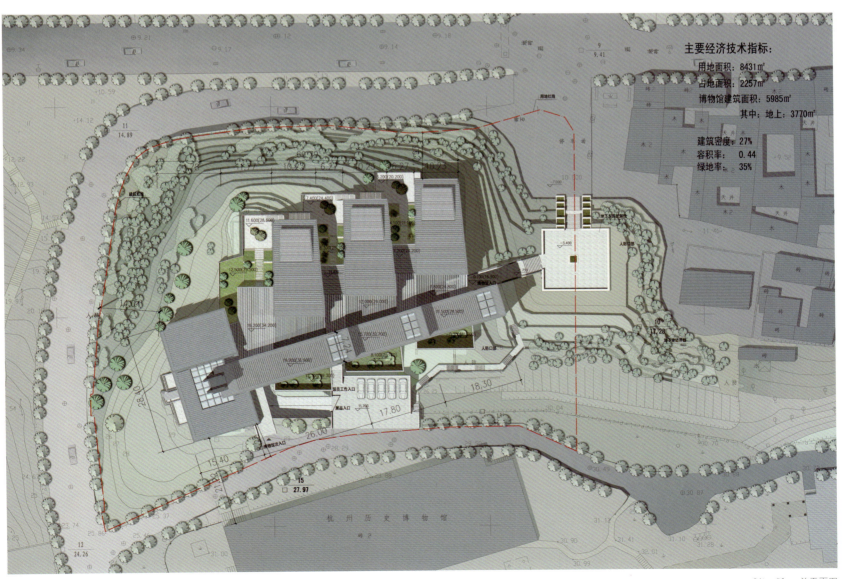

主要经济技术指标:
用地面积: 8431㎡
占地面积: 2257㎡
博物馆建筑面积: 5985㎡
其中: 地上: 3770㎡
建筑密度: 27%
容积率: 0.44
绿地率: 35%

Site Plan 总平面图

North Elevation 北立面图

Architectural Design

The design makes full use of geographical advantages, and integrates the relationship among site, surrounding buildings and natural environment to improve the quality of the whole environment. A building that possesses the characteristics of museum and contains public and open space is created, providing citizens a good place for cultural activities. An architectural space form integrates with natural environment and cultural environment is organized; contemporary architecture image with local traditional image is created.

The building design connects some exhibition spaces with finger layout through the walking spindle that extends along the east-west contour, the building complies with the landform, and develops into continuous and rhythmic spatial pattern. Additional courtyard and the terrace make the building and natural environment permeate and blend each other.

建筑设计

设计充分利用区位优势，整合地块与周边建筑及自然环境的关系，提升地块的整体环境品质；创造既有博物馆建筑性格特点，又具有公共性和开放性的场所空间，使之成为市民文化活动的良好去处；组织与自然环境、人文环境相互融合的建筑空间形态，创造当代且兼具地方传统意象的建筑形象。

建筑由沿等高线东西向展开的步行主轴，将若干指状布局的展示空间联系起来，并顺应山势，逐步抬高，形成整体连续、张弛有度的空间格局。通过穿插其间的庭院和观景平台使自然环境和建筑完美整合。

KEY WORDS	Stone Facade 石材外墙
关键词	Sculptural Sensibility 雕塑感
	Artistic Atmosphere 艺术氛围

China National Theatre and the Office
中国国家话剧院剧场及办公楼

FEATURES 项目亮点

The shape of the building drew inspiration from Chinese traditional architecture and stage elements, expresses unique traditional architectural elegance via dignified appearance and exquisite details.

建筑造型设计从中国传统建筑和舞台元素中汲取灵感,以平和、厚重的外观和精致、典雅的细节彰显出建筑特有的韵味。

Location: Xuanwu District, Beijing, China
Architectural Design: Chinese Aviation Planning and Construction Development Co., Ltd
Land Area: 10,166.23 m²
Total Floor Area: 21,000 m² (Theater 15,000 m², Office 6,000 m²)
Completion: 2011

Overview

China National Theater is located in northeastern corner of the intersection between Guang'anmen Avenue and Sanli Henanyan Road, Xuanwu District, Beijing. The project includes 888 professional theater and the office buildings; the theater design meets drama performing requirements of national grade A theatre. After completed, China National Theater provides a permanent art practical base.

项目地点:中国北京市宣武区
设计单位:中国航空规划建设发展有限公司
用地面积:10 166.23 m²
总建筑面积:21 000 m²(剧场15 000 m²,办公楼6 000 m²)
竣工时间:2011年

项目概况

中国国家话剧院位于北京宣武区广安门外大街与三里河南延路交叉口的东北角。建筑包括888座专业剧场和业务办公楼两部分,其中设计了满足话剧演出要求的国家甲等剧场。建筑的落成为中国国家话剧院提供了永久的艺术实践基地。

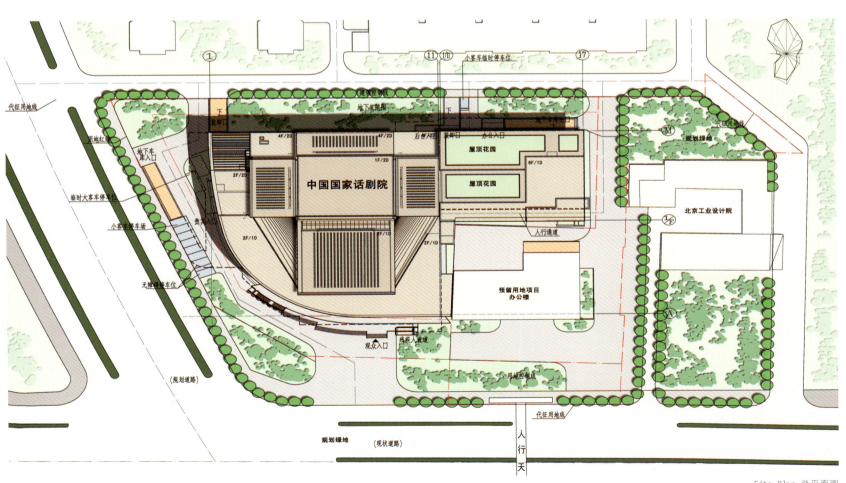

Site Plan 总平面图

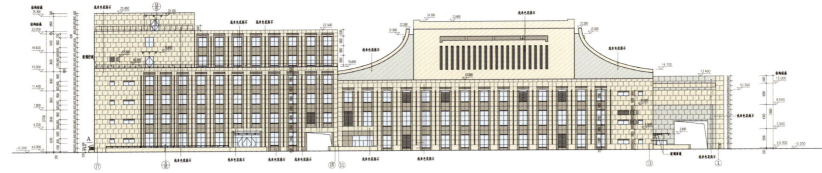

North Elevation 北立面图

South Elevation 南立面图

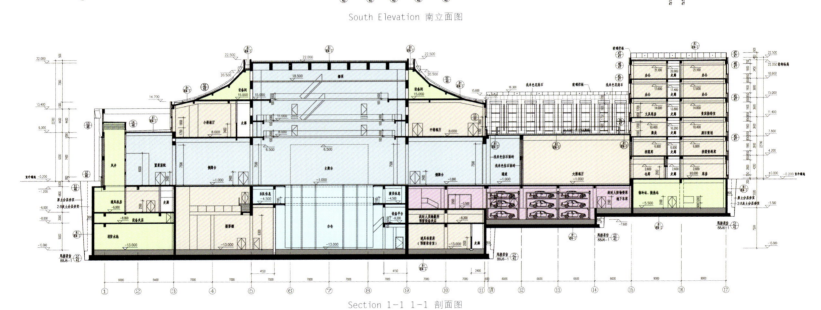

Section 1-1 1-1 剖面图

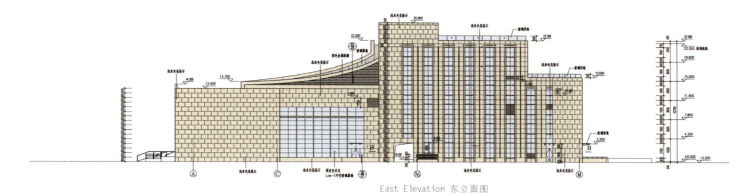
East Elevation 东立面图

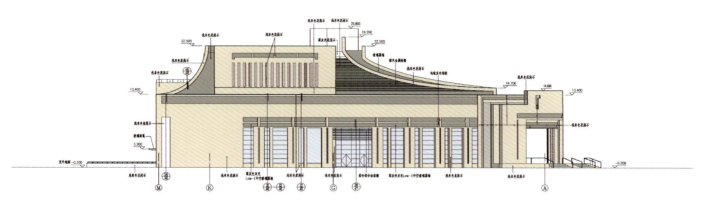
West Elevation 西立面图

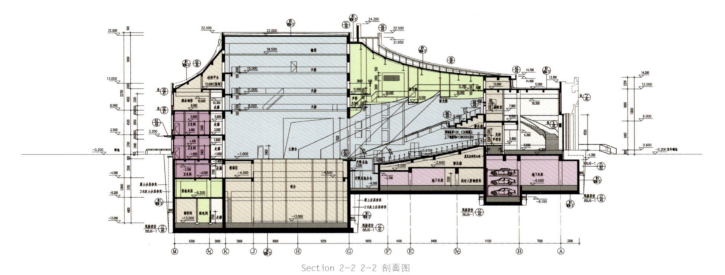
Section 2-2 2-2 剖面图

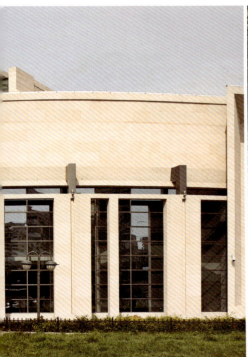

First Floor Plan 一层平面图

Second Floor Plan 二层平面图

Plan for Basement Floor 地下一层平面图

Architectural Design

The shape of the building drew inspiration from Chinese traditional architecture and stage elements. Taking advantage of the higher section of the tower, lofty and powerful hipped roof are shaped, which not only reflects the traditional architectural elegance, but also highlights the memorial character of the building. The main entrance is like a huge "entablature", expressing the artistic theme of the theater and giving the audiences a feeling of stepping into the art palace. The building shape is combined with site contour, it is very flexible within this limited site, and properly expresses the public nature of cultural building. The building shape is dignified, with strong sense of emptiness and reality. Stone as the main decoration for facade, with strong sculptural sensibility, and reflects the profound cultural accumulation. Architectural details are exquisite, double colonnade with rich level and sense of sequence are endowed. The shadow of the entrance with distinct relationship level, to enhance sculptural sensibility, the exquisite eavescornice reflects the sense of decoration. The main foyer with its arc shape develops into stretched public space; a large staircase connects the lobby on the first and second floor, the spatial variation is rich; the ceiling at the top stretched from the tapered fan emphasizes spatial continuity; the bronze themed column highlights the drama's artistic theme and creates strong artistic atmosphere.

建筑设计

造型设计从中国传统建筑和舞台元素中汲取灵感。利用台塔的较高部分形成巍峨、雄浑的四坡顶，既体现传统建筑韵味，又凸显建筑纪念性。建筑主入口形如巨大的"台口"，表现舞台艺术主题，给观众以步入艺术殿堂的意境。建筑外形结合基地轮廓，在有限用地内张弛有度，恰当地表现文化建筑的公共性。建筑造型体量厚重，虚实感强。外墙装饰以石材为主，雕塑感强，彰显出深厚的文化积淀。建筑细节刻画细腻，双柱柱廊层次丰富，赋予序列感。入口阴影关系层次分明，极具雕塑感，檐口线脚细腻，体现装饰感。大堂利用弧形形成舒展的公共空间，大型楼梯连接一、二层休息厅，空间变化丰富；顶部吊顶由锥状单元扇形展开，强调空间连续性；青铜主题柱凸显话剧艺术主题，营造浓郁的艺术氛围。

KEY WORDS 关键词

Glass Facade 玻璃立面

Unique Appearance 外观独特

Checkered with Sunlight and Shade 光影交幻

Chongqing Science & Technology Museum
重庆科技馆

FEATURES 项目亮点

The facade design expresses the concept of scenic town, compared stone with mountain, glass as water, the stone with different shades of color, and its texture with thickness and thin interlaced, simulating the shape of majestic rocks in the nature. Transparent glass directly inserted into stone creates the effect checkered with sunlight and shade.

建筑立面结合山水之城的理念，以石材喻山、玻璃喻水，石材上颜色深浅不一，质地粗细交错，模拟自然界中巍峨岩石的形态。由透明玻璃构成的部分建筑直接插入石材构成的主体建筑，形成光影交幻的效果。

Location: Jiangbei District, Chongqing, China
Architectural Design: Chongqing Design Institute
Land Aea: 24,764 m²
Total Floor Area: 40,302 m²
Plot Ratio: 1.63

项目地点：中国重庆市江北区
设计单位：重庆市设计院
用地面积：24 764 m²
总建筑面积：40 302 m²
容积率：1.63

Planning and Layout

The building is divided into main building and wing building according to function. All kinds of exhibition halls and giant-screen cinema are arranged on the main building, the wing building satisfies functional requirements of conference, training, and office, it also develops into parking lot, small shop, equipment space and other auxiliary spaces through combining lower part of landscape main axle square and the basement.

规划布局

建筑平面按功能分为主楼和附楼两部分。主楼设置各类展厅和巨幕影院；附楼除满足会议交流、培训、办公等功能的需求外，还结合景观主轴广场下部空间及本身的地下室设置了各类停车场、小型商业、设备用房等辅助空间。

Site Plan 总平面图

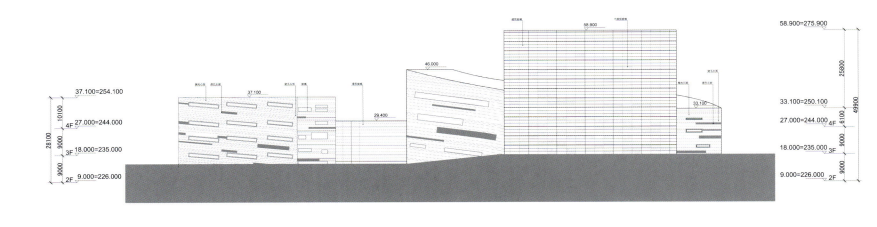

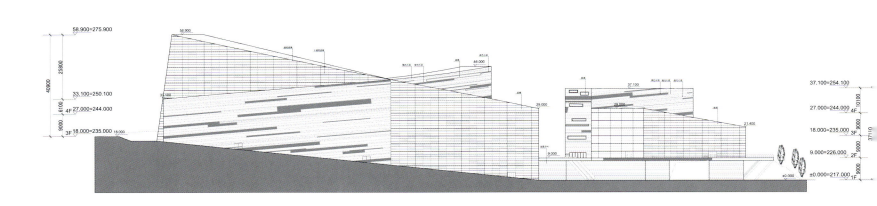

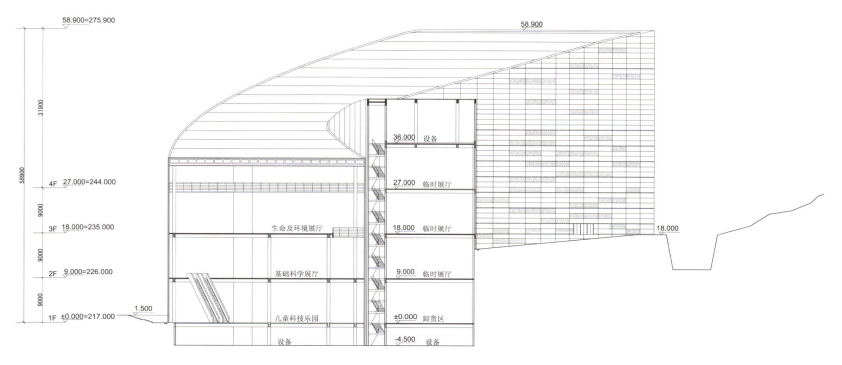

Sectional Drawing 剖面图

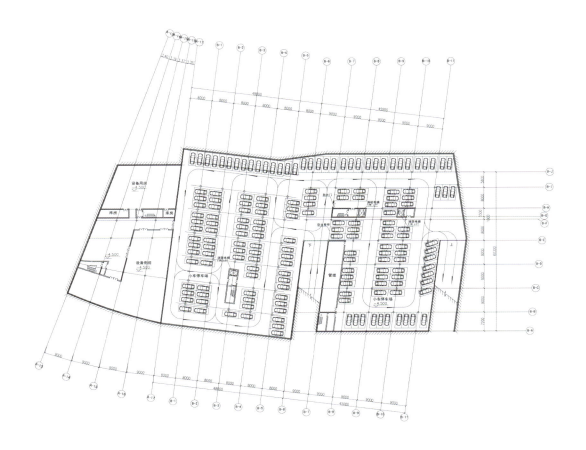

Architectural Design

The facade design expresses the concept of scenic town, compared stone with mountain, glass as water, the stone with different shades of color, and its texture with thickness and thin interlaced, simulating the shape of majestic rocks in the nature. The arched and corner angle are distinct; the arc is like water and the corner angle is like mountain; the building appearance is concise with fashionable style, and contains the characteristics of scenic Chongqing. Transparent glass directly inserted into stone creates the effect checkered with sunlight and shade. The arched flank reflects local topographic characteristics, and runs through parallel exhibition spaces; through expression and interpretation of these two materials, it expresses primitive simplicity but modern Chongqing style, to work in concert with the architectural style of Jiefanggbei.

建筑设计

建筑立面结合山水之城的理念，以石材喻山、玻璃喻水，石材上颜色深浅不一，质地粗细交错，模拟自然界中巍峨岩石的形态。线条棱角分明，弧形似水，棱角如山，外形简洁，风格前卫，尽显重庆山水之城的地域特征。由透明玻璃构成的部分建筑直接插入石材构成的主体建筑，形成光影交幻的效果。而弓形的侧面体现当地的地形特点，并将各层平行展区连为一体，通过对这两种材料的展现和解读，彰显出古朴而现代的"重庆风格"，以此呼应解放碑的建筑风格。

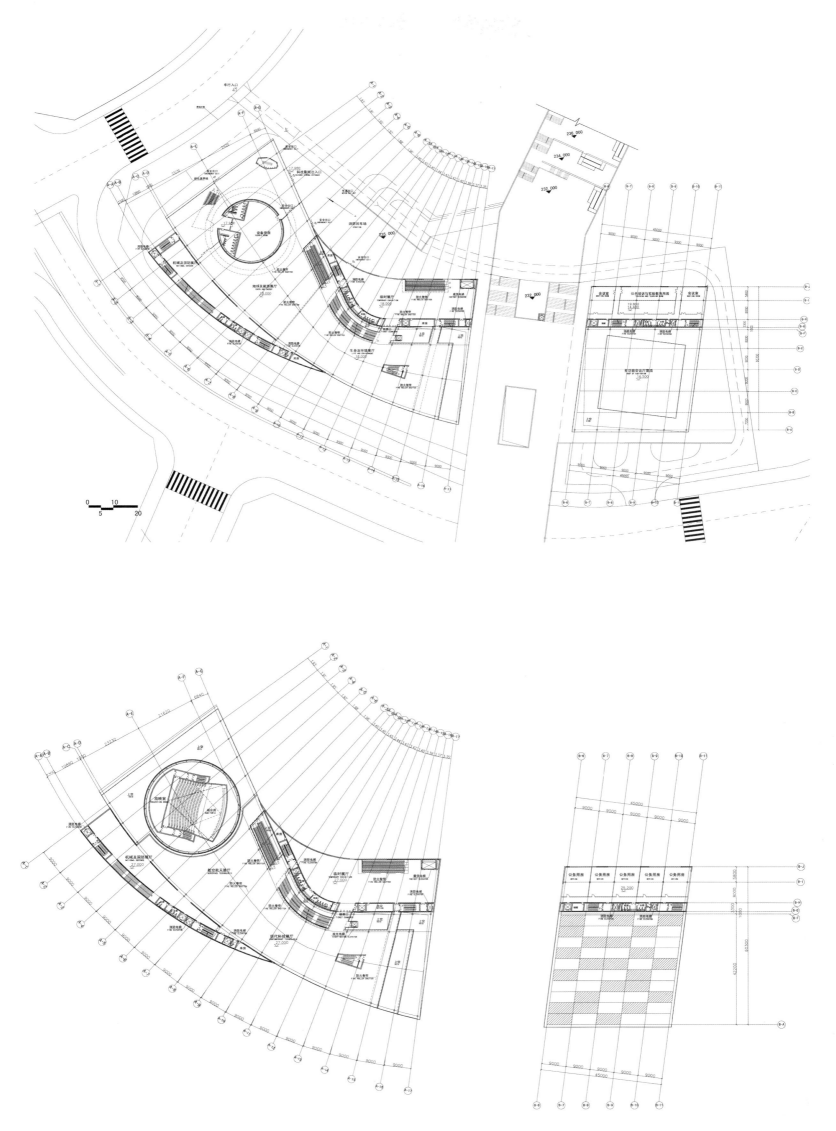

KEY WORDS 关键词

Petal Shape 花瓣造型
Glass Facade 玻璃立面
Urban Landscape 城市景观

Xuzhou Music Hall
徐州音乐厅

FEATURES 项目亮点

The design draws inspiration form the shape of crape myrtle flower, with abstract architectural techniques; the overall building is like the spreading petal layer by layer, blossoming on the tranquility water surface of Yulong Lake.

音乐厅设计取意紫薇花的形态，采用抽象的建筑手法，整体犹如层层展开的花瓣，绽放在云龙湖平静的水面上。

Location: Xuzhou, jiangsu, China
Architectural Design: Tsinghua University Architecture Design and Research Institute Co., Ltd
Total Land Area: about 40, 000 m²
Total Floor Area: 13,322 m²
Building Height: 28.9 m
Plot Ratio: 0.33
Green Coverage Ratio: 45.6%
Completion: 2011

项目地点：中国江苏省徐州市
设计单位：清华大学建筑设计研究院有限公司
总用地面积：约 40 000 m²
总建筑面积：13 322 m²
建筑高度：28.9 m
容积率：0.33
绿地率：45.6%
竣工时间：2011 年

Overview

Xuzhou Music Hall is located in northern shore of Yunlong Lake, Xuzhou city, the project is constituted by hall, outdoor performance square, hall entrance plaza, with a total floor area of 13,322 m². The music hall with the floor area of 10,530 m², and it can accommodate 986 spectators.

项目概况

徐州音乐厅位于徐州市云龙湖北岸，项目包括音乐厅、室外演出广场、音乐厅入口广场三部分，总建筑面积 13 322 m²。其中，音乐厅建筑面积 10 530 m²，可容纳 986 名观众。

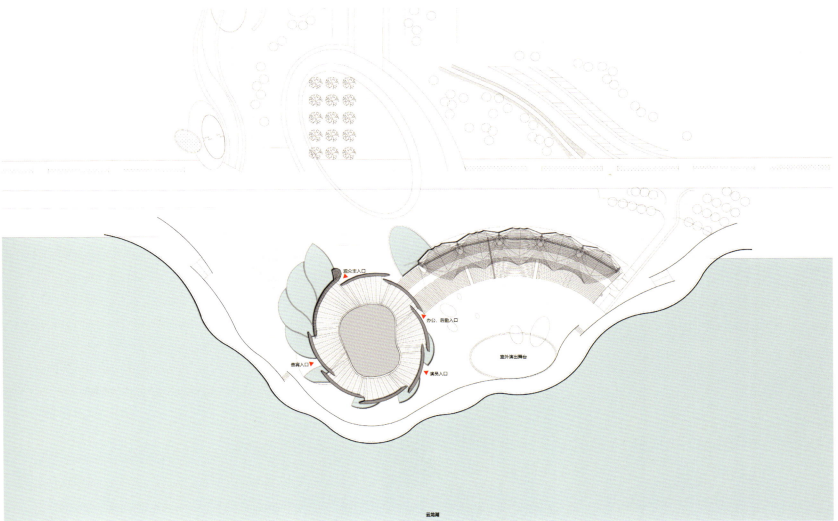

Site Plan 总平面图

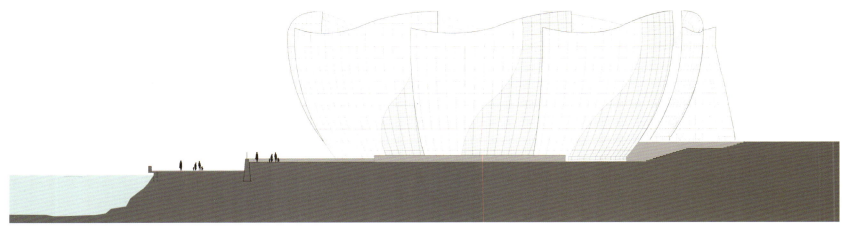

East Elevation 东立面图

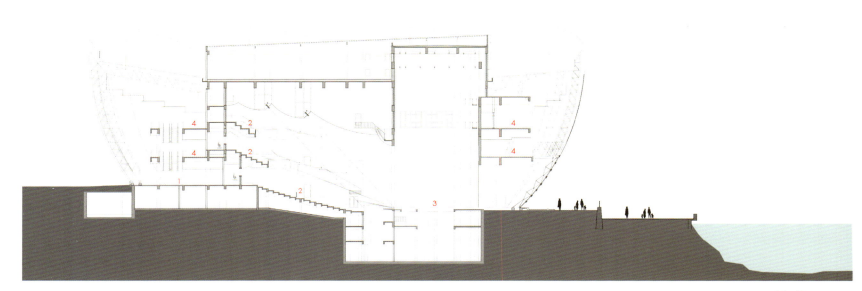

Sectional Drawing 剖面图

1. 入口大厅
2. 观众厅
3. 舞台
4. 观景平台

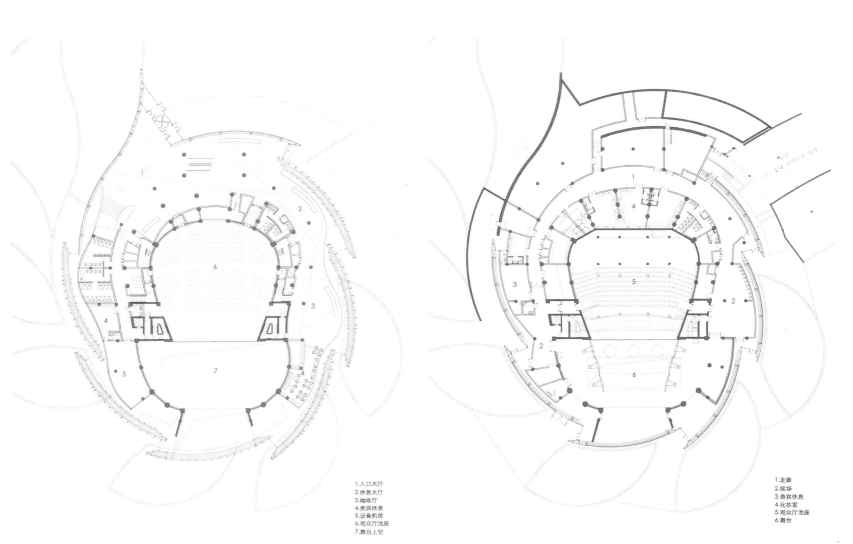

1. 入口大厅
2. 休息大厅
3. 咖啡厅
4. 贵宾休息
5. 设备机房
6. 观众厅池座
7. 舞台上空

1. 走廊
2. 候场
3. 贵宾休息
4. 化妆室
5. 观众厅池座
6. 舞台

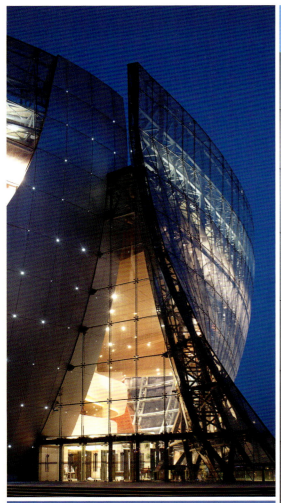

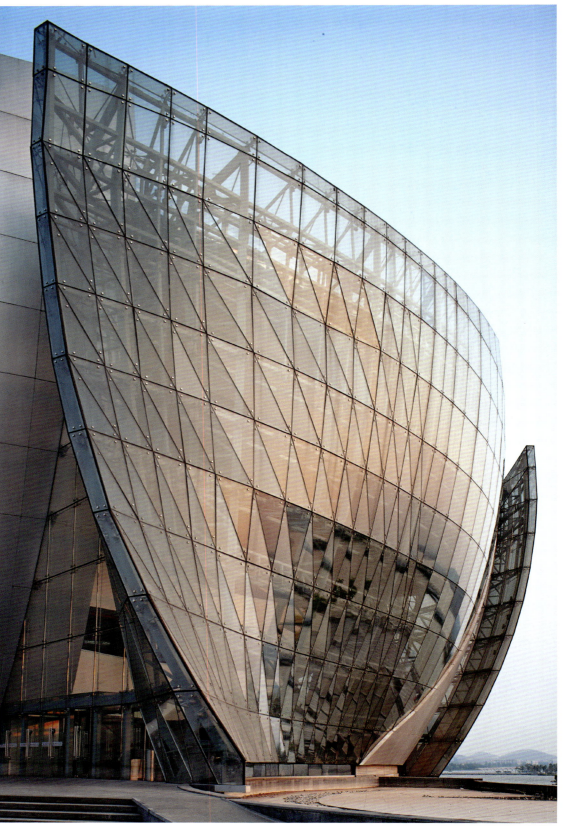

Architectural Design

Xuzhou Music Hall is built on a lake peninsula developed by artificial fill lake, the design draws inspiration form city's flower, the shape of crape myrtle flower. The overall building is like the spreading petal layer by layer, draws graceful shape of the flower, blossoming on the tranquility water surface of Yulong Lake. The main body is surrounded by open outdoor stands and recreational space, regarded the beautiful Yunlong Lake as stage background, after completed, it becomes an open stage for urban public performances, and provides citizens leisure public space to overview the lake. The building conforms to location, topography and functional characteristics, develops into rational and symbolic architecture form, integrates with urban landscape, and symbolizes the enterprising spirit of the city.

建筑设计

徐州音乐厅建造在一片人工填湖形成的临湖半岛用地上，设计取意城市的市花——紫薇花的形态。建筑整体犹如层层展开的花瓣，勾勒出婀娜的形态，绽放在云龙湖平静的水面上。开放的室外看台及休闲空间环绕建筑主体，以美丽的云龙湖为舞台背景，成为城市公共演出的开放舞台，也为市民提供一处观湖休闲的公共空间。建筑契合场所、地形、功能特征，形成具有象征性的理性的建筑形态，融入山水城市景观，象征着进取的城市精神。

KEY WORDS 关键词

Frame Structure 框架结构
Roof System 屋面结构
Glass Curtain Wall 玻璃幕墙

Urban Planning Exhibition Hall, Liu'an, Anhui
安徽省六安市城市规划展览馆

FEATURES 项目亮点

The cubic office building shapes a contrast with the oval shaped exhibition hall. The eaves rise gradually around the facades, forming a unique outline for the building. The skin is covered by black slate curtain wall, creating an integrated and rhythmic appearance.

办公区简洁的矩形体量与椭圆形的展厅形成鲜明而极具张力的对比，沿立面四周逐渐升起的屋檐，使简单的形体产生微妙的轮廓变化；屋檐之下的建筑表皮，是连续的黑色板岩幕墙，赋予建筑外观视觉上的整体性和韵律感。

Location: Liu'an, Anhui, China
Architectural Design: Jianxue Architecture and Engineering Design Institute Co.,Ltd.
Total Land Area: 22,000 m²
Total Floor Area: 11,119.08 m²

项目地点：中国安徽省六安市
设计单位：建学建筑与工程设计所有限公司
总用地面积：22 000 m²
总建筑面积：11 119.08 m²

Overview

The exhibition hall is situated in the new administrative district of Liu'an City, Anhui Province, at the southwest corner of the municipal administrative center. It is close to Foziling Road in the south with an exhibition area of 6,400 m² and a total office area of 4,700 m².

项目概况

六安市城市规划展览馆位于安徽省六安市南部政务新区内，在市行政中心西南角，南临佛子岭路。展览面积6 400 m²，附属办公用房4 700 m²。

Planning

The site continues the gentle slope terrain of the administration center to be lower in the south and higher in the north, with an altitude difference of 5 m. The architects take advantage of this kind of terrain and set the exhibition space in the south which is covered by vegetation roof. The green roof extends to the north and connects with the sloping field. The building houses the offices and warehouses, the equipment rooms are built in the north and connected with the exhibition space at the bottom.

The entrance of the exhibition hall opens to the urban road, while the office building's entrance is set in behind on the slope to connect with the internal ring road. All the functions are well arranged, and the building looks elegant to dialogue with its surroundings.

规划布局

建筑基地延续了行政中心的丘陵缓坡地形，南低北高，高差近5m。总体布局充分结合地形，将展览空间置于基地南侧较低处，其上以整体的植草屋面覆盖，向北延伸与坡地直接相连。办公及辅助用房则置于基地北侧，其上部为办公用房，底部为库房及设备技术用房并与展览空间相连。

展厅出入口面向两侧城市道路，办公出入口位于基地后方坡地上，经由沿用地边线环路进出。总体设计一气呵成，功能分区合理、明确，流线组织清晰、便捷，建筑形体新颖、简洁。建筑与场地实现了对接。

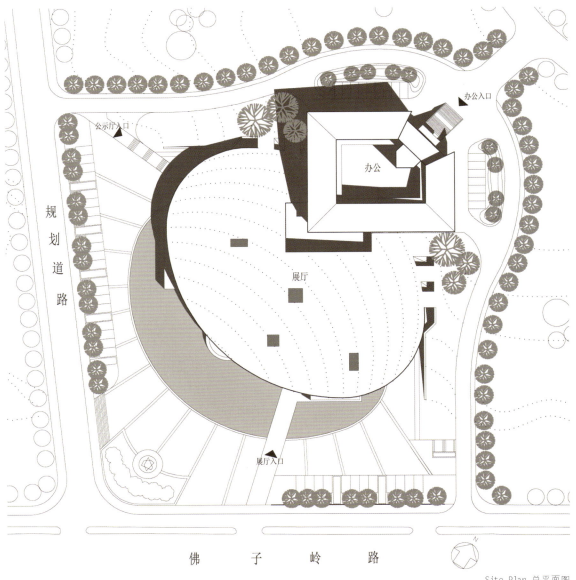

Site Plan 总平面图

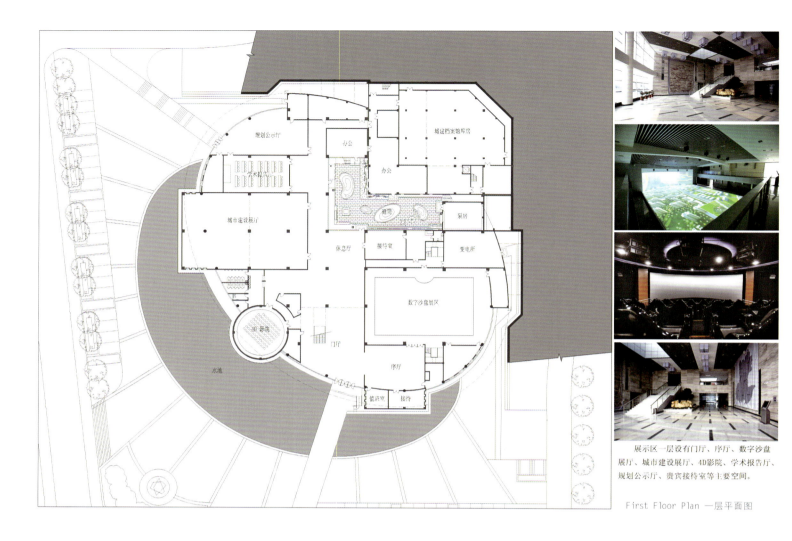

展示区一层设有门厅、序厅、数字沙盘展厅、城市建设展厅、4D影院、学术报告厅、规划公示厅、贵宾接待室等主要空间。

First Floor Plan 一层平面图

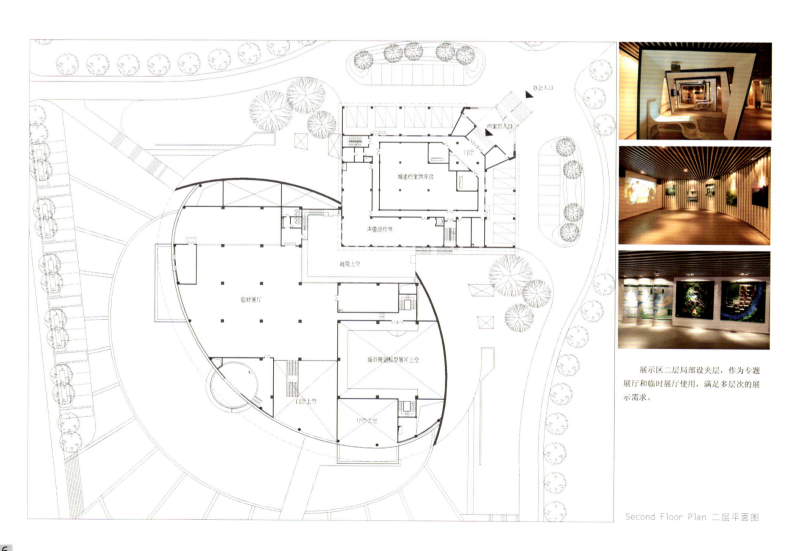

展示区二层局部设夹层，作为专题展厅和临时展厅使用，满足多层次的展示需求。

Second Floor Plan 二层平面图

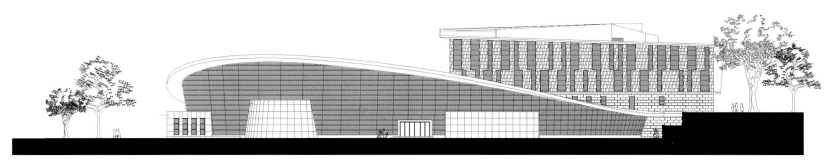

South Elevation 南立面图

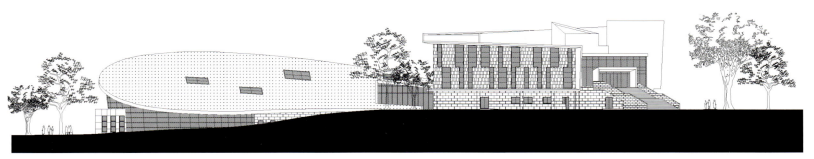

East Elevation 东立面图

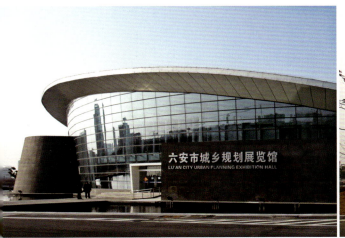

Roof System

With the limitation of the local construction techniques and the control of investment, the architects have persuaded the client to undertake the exhibition design and architectural design at the same time. Under the double curved roof, concrete frame structure with customized spaces is designed instead of the large space structure in common exhibition buildings. 3D software is used to complete the innovative design.

It is the biggest grass roof in this area. Both of its appearance and utilization agree to the expected design effect.

Shape of the Building Volume

The projecting eave draws a beautiful arc which feels like the portal to the city. Glass curtain wall slightly tilts inwards and witnesses the changes of the city under the sunshine and on the reflection of the pool. In the exhibition area, huge oval roof covered by grass has continued the undulating terrain behind the site, looking like a lifted green land. Beneath it are exhibition spaces of different sized to embrace the history, presenting and future of the city.

The cubic office building shapes a contrast with the oval shaped exhibition hall. The eaves rise gradually around the facades, forming a unique outline for the building. The skin is covered by black slate curtain wall, creating an integrated and rhythmic appearance. Skillful design makes this kind of cheap and common material more impressive.

屋面结构

考虑到当地施工技术条件限制和投资控制等因素，覆盖整个展示区的双曲面屋面并未选用展览建筑常用的大空间结构，而是在方案阶段说服建设单位引入展览策划机构与建造设计同步进行，根据各展厅的空间尺度要求，采用普通的混凝土框架结构，化大为小，确定合理的结构柱网，并借助于三维信息化设计软件精心设计。

屋面为当地最大的植草屋面，从建成后的外观和使用效果来看，均较好地实现了设计的预期效果。

建筑造型

深远的挑檐勾勒出一道弧线，好像开启了城市时空之门，微微向内倾斜的光洁的玻璃幕墙在阳光和水池的映射中，见证着城市的变迁。展示区中巨大的椭圆形植草屋面延续了基地后部自然的丘陵地貌，缓缓地伸向城市天空，仿佛是掀起的一片绿地，而覆盖在它下面尺度各异的展示空间，则浓缩了城市的历史、现在和未来。

办公区简洁的矩形体量与椭圆形的展厅形成鲜明而极具张力的对比。沿立面四周逐渐升起的屋檐，使简单的形体产生微妙的轮廓变化；屋檐之下的建筑表皮，是连续的黑色板岩幕墙，赋予建筑外观视觉上的整体性和韵律感。精巧而仔细的设计，使板岩这种廉价而常见的天然材料彰显出独特的艺术表现力。

KEY WORDS 关键词

- **Complete Function** 功能完备
- **Innovative Design** 造型新颖
- **Convenient Traffic** 交通便利

Beijing International Flower Logistic Port
北京国际花卉物流港

FEATURES 项目亮点

Designers took 36 flowers umbrellas and artificial vegetation leaves as the basic shape for the port, which fully embodied the innovative energy-saving and practical design philosophy, and achieved a perfect combination of technology and art, no wonder it was another landmark for Beijing after Beijing Olympic Venues.

北京国际花卉物流港以 36 个花伞和仿生植物生长叶为基本造型，充分体现了绿色节能实用的新颖设计理念，是技术与艺术的完美结合，是继奥运场馆之后北京的又一标志性建筑。

Location: Shunyi, Beijing, China
Architectural Design: An-design Architects
Building Scale: 166,492 m² (36,124 m² underground)
Land Area: Phase I 138,216 m², Phase II 90,000 m²
Building Height: 44 m
Plot Ratio: 1.15

项目地点：中国北京市顺义区
设计单位：北京华清安地建筑设计事务所有限公司
建筑规模：166 492 m²（地下为 36 123 m²）
用地面积：一期为 138 216 m²，二期为 90 000 m²
建筑高度：44 m
容积率：1.15

Overview

This project consists of 1#, 2#, 3# exhibition halls, a logistics center and a trading center. Boasting 6 floors aboveground and 1 floor underground, it was the largest professional site that combines production, R&D, exposition and trade.

项目概况

项目由 1、2、3 号展馆，物流中心及交易中心组成。地上 6 层，地下 1 层。它是国内迄今为止最大的集花卉展览、物流、研发、交易于一体的专业性花卉场所。

Planning

Compact layout was used to save the land. North-south planning road passed through the site and divided it into two parts: the east side and the west side. During the Expo, the main buildings such as the 1#, 2#, 3# exhibition halls, the logistics center and the trading center were placed on the east side, and exhibition space was arranged in two floors and the office space was arranged above the exhibition area to reduce construction area. While the west side was used for temporary ground parking and exhibition greenhouse, and later for hotel, apartments, etc. after the Expo.

规划布局

项目采用紧凑式布局实现节地，利用穿过场地内部的南北向规划道路将场地分为东西两块。将花博会期间主要建筑 1、2、3 号展馆，物流中心和交易中心都布置在场地东侧，且将展览空间布置为两层，办公空间置于展览空间之上，减少建筑占地面积。西侧场地在会时用作临时地面停车及展销大棚，会后则为酒店、配套公寓和远期开发用地。

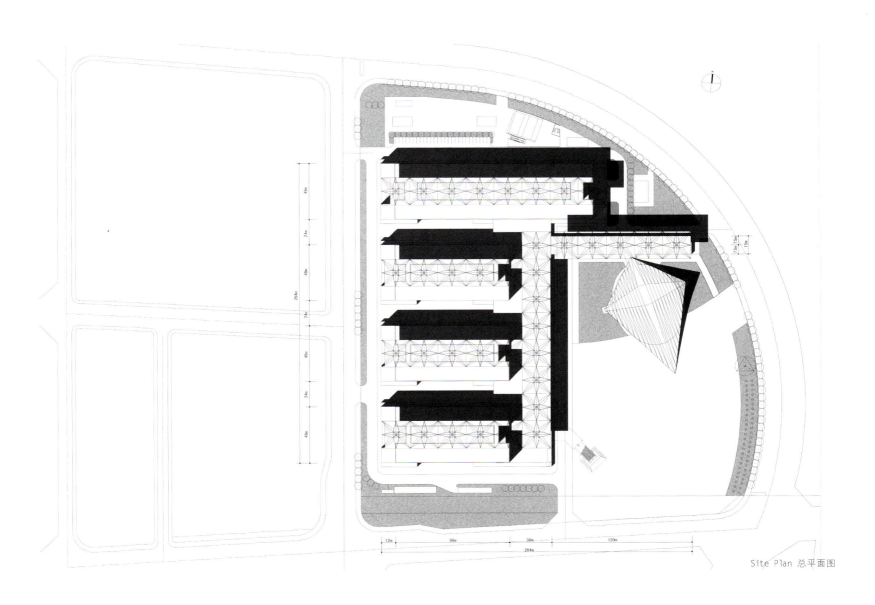

Site Plan 总平面图

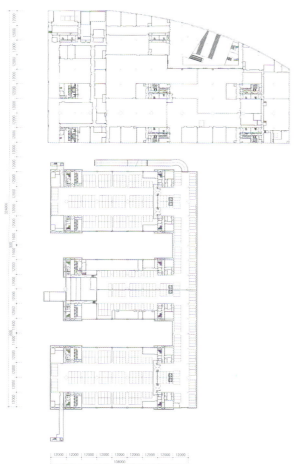

Plan for Basement Floor 地下一层平面图

First Floor Plan 一层平面图

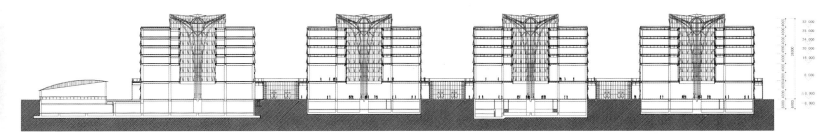

Section 1-1 1-1 剖面图

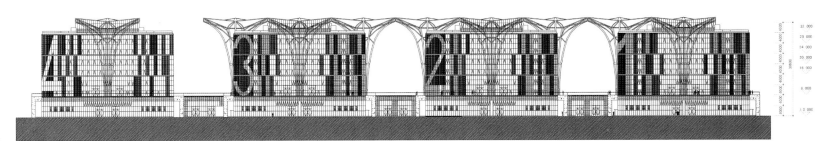

West Elevation 西立面图

Architectural Design

Designers took 36 flowers umbrellas and artificial vegetation leaves as the basic shape for the port, which fully embodied the innovative energy-saving and practical design philosophy, and achieved a perfect combination of technology and art, no wonder it was another landmark for Beijing after Beijing Olympic venues.

Though it was built for the 7th China Flowers Exposition, the spatial scale it presented was far beyond that single function. Two-storey exhibition space enjoyed an average 8 m on each floor, a required height that met all the needs to exhibit flowers during the Expo. After the Expo, the floors in that height can also be used for a variety of large commercial exhibitions.

建筑设计

北京国际花卉物流港以36个花伞和仿生植物生长叶为基本造型，充分体现了绿色节能实用的新颖设计理念，是技术与艺术的完美结合，是继奥运场馆之后北京的又一标志性建筑。

北京国际花卉物流港虽为第七届中国花卉博览会而建，但它所呈现的空间尺度却远远超出了这一单一功能。花博会期间用以展览花卉的展馆一层和二层，层高均为8米，在花博会期间可满足展览花卉所需的高度。花博会之后，这样的层高同样可以用于举办各种大型商业展销活动。

Traffic Flow Line

Major flow was gathered on the north-south planning road and VIP entrance was opened in the middle of the south side road of Northern Airport Expressway to the east of the site. The southeast of the site was a public square that opened to the city for pedestrians during the Expo.

交通流线

会时的车流主要在南北向的规划路上解决，而在场地东侧的机场北线南辅路中段开辟了贵宾车行出入口。场地东南部分为公共广场，向城市打开，为会时主要的人行活动场所。

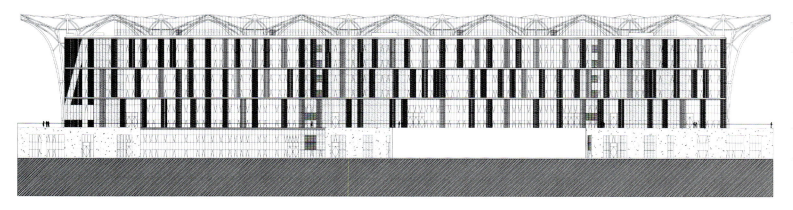

North Elevation 北立面图

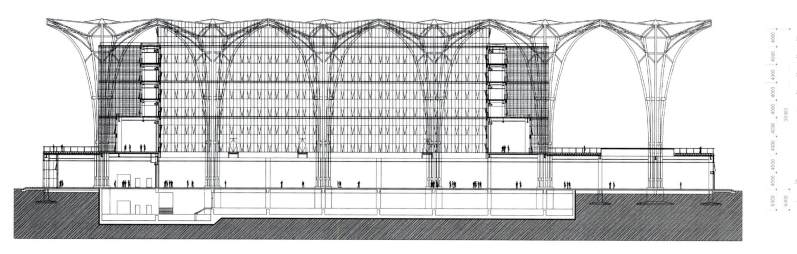

Section 2-2 2-2 剖面图

KEY WORDS 关键词

Arc-shaped Wall Space 弧形墙面
Glass Curtain Wall 玻璃幕墙
Arc-shaped Portico 弧形柱廊

Hebei Qian'an Cultural Convention and Exhibition Center
河北迁安市文化会展中心

FEATURES 项目亮点

The east-west axis in the People's Square is the major axis for the entire project. 18 m arc-shaped wall space opposite the People's Square echoes with the arc-shaped portico on the Square, forming an enclosure to the square together; and they are viewed as the visual center and the highest point.

整个建筑以人民广场的东西轴线为主轴，面向人民广场方向为18 m高的弧形墙面，和广场的弧形柱廊相呼应，共同形成对广场的围合，也是广场的视觉中心和最高点。

Location: Tangshan, Hebei, China
Architectural Design: An-design Architects
Total Floor Area: 25,025 m²

项目地点：中国河北省唐山市
设计单位：北京华清安地建筑设计事务所有限公司
总建筑面积：25 025 m²

Overview

The Cultural Convention and Exhibition Center is comprised of three parts: convention center, library and cultural activities center. The total floor area is 25,025 m², in which 5,138 m² is the 4-storey cultural activities center and 14,749 m² is for the convention center which has varied floors (1 to 3) for different parts.

项目概况

文化会展中心由会展中心、图书馆和文化活动中心三部分组成。总建筑面积为25 025 m²，文化活动中心部分为5 138 m²，层数为四层；会展中心部分为14 749 m²，层数局部为三层和两层，局部为一层。

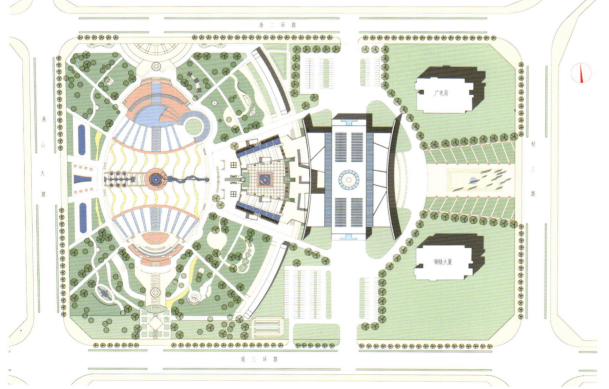

Site Plan 总平面图

Architectural Design

The east-west axis in the People's Square is the major axis for the entire project. 18 m arc-shaped wall space opposite the People's Square echoes with the arc-shaped portico on the Square, forming an enclosure to the square together and they are viewed as the visual center and the highest point. In the center of arc-shaped wall space is a 15 m high gate; the sculptures that reflect historic culture of Qian'an exhibited on the outdoor hallway is known as the gate of history. Passing through the gate, one can see the sparkling point supported glass curtain wall of the convention center, behind the curtain wall there is a huge spherical dome screen which is 27 m in diameter, like a bright pearl and a rising red sun, indicating the bright future of Qian'an. The roof is also in arc shape, like the royal crown of Huangdi (Yellow Emperor), conveying the past of Qian'an. The east façade is a 20 m high exposed framing glass curtain wall. And the east gate represents the fast-changing of technology in the information era, known as the gate of the era.

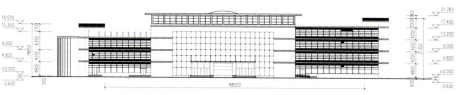

Elevation 1 立面图 1

Elevation 2 立面图 2

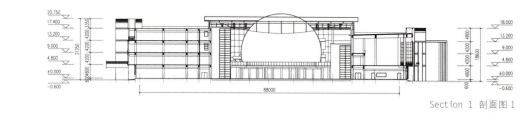

Section 1 剖面图 1

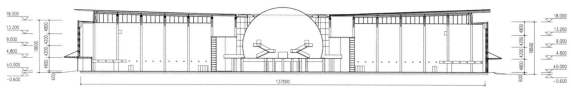

Section 2 剖面图 2

建筑设计

整个建筑以人民广场的东西轴线为主轴，面向人民广场方向为 18 m 高的弧形墙面，和广场的弧形柱廊相呼应，共同形成对广场的围合，也是广场的视觉中心和最高点。弧形墙面的中央是一个 15 m 高的大门，露天的门厅内摆设反映迁安历史文化的雕塑，被称为"历史之门"。穿过"历史之门"，会看到会展中心的点式玻璃幕墙熠熠生辉，幕墙背后映入眼帘的是一个直径 27 m 的巨大球形穹幕，像一颗璀璨的明珠，又像一轮升起的红日，预示着迁安腾飞的未来，并和两个弧形墙面形成双龙拱日之态。屋顶部分呈两侧向上的弧面造型，纵贯南北，仿佛黄帝的冠冕，彰显出迁安作为黄帝故都的文脉，东侧主立面是一个高达 20 m 的明框玻璃幕墙，东大门代表着信息时代科技的日新月异，被称为"时代之门"。

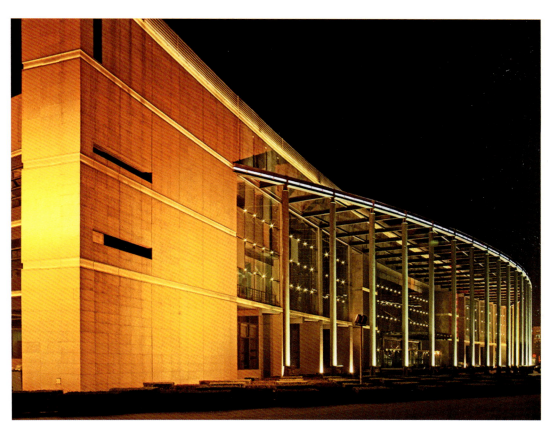

KEY WORDS 关键词

Long Shape 形体较长

Elegant Appearance 外观大气

Green System 绿化系统

Longyan Convention and Exhibition Center
龙岩市会展中心

FEATURES 项目亮点

This long-shaped building features some openings and has varied spaces interwoven with each other to optimize the meeting spaces and present convention spaces of new style.

空间层次丰富，根据建筑形体较长的特点，在适当部位进行开口处理，各种空间相互穿插、流动，极大地改善了会议空间的自然条件，创造了全新的建筑空间形态。

Location: Longyan, Fujian, China
Architectural Design: Fujian Provincial Institute of Architectural Design and Research
Total Land Area: 115,071.47 m²
Total Floor Area: 40,458.39 m²
Construction Area: 11,104.1 m²
Building Density: 9.6%
Plot Ratio: 0.26
Green Coverage Ratio: 30.6%

项目地点：中国福建龙岩市
设计单位：福建省建筑设计研究院
总用地面积：115 071.47 m²
总建筑面积：40 458.39 m²
建筑基底面积：11 104.1 m²
建筑密度：9.6%
容积率：0.26
绿地率：30.6%

Architectural Design

Taking advantage of the elevation different, different floors are designed with different meeting rooms to create a series of staggered and dynamic spaces for meeting and recreation. This long-shaped building features some openings and has varied spaces interwoven with each other. This kind of design has greatly optimized the meeting spaces and presented a new-style convention building. And the multi-level green system has helped to create a beautiful and comfortable environment.

The design shows respect to the local culture and topography to preserve the existing trees and vegetations, and set a Banyan Tree Square in the convention center.

建筑设计

按楼层布置不同的会议空间，通过对场地高差的处理，形成相互错动、充满动感的会议与休闲空间。空间层次丰富，根据建筑形体较长的特点，在适当部位进行开口处理，各种空间相互穿插、流动，极大地改善了会议空间的自然条件，创造了全新的建筑空间形态。多标高、多层次的绿化系统，创造舒适、优美的会议环境。

设计尊重当地风土人情，尊重地形地貌。充分保留了基地内原有古树名木及原生植被，结合古榕树在会议中心布置榕树广场。

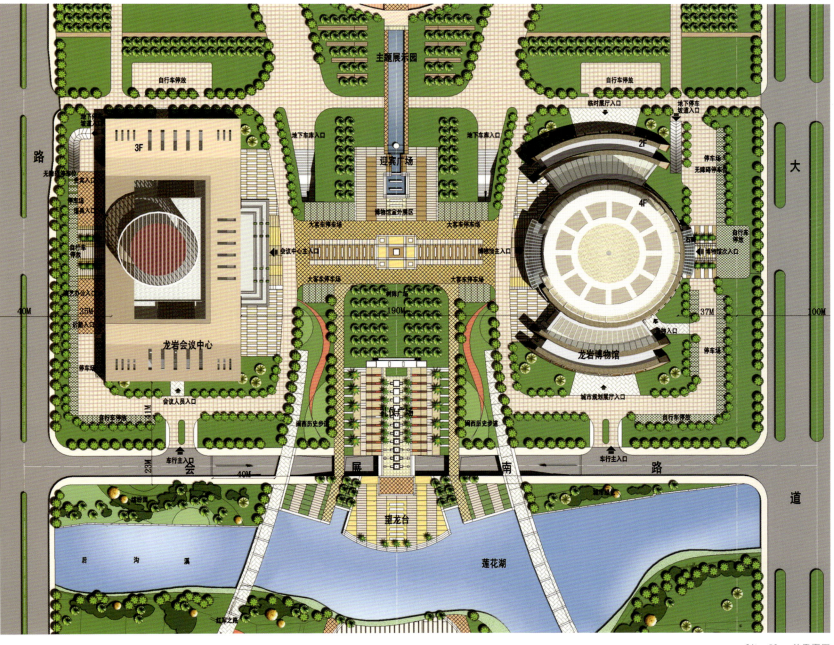

Site Plan 总平面图

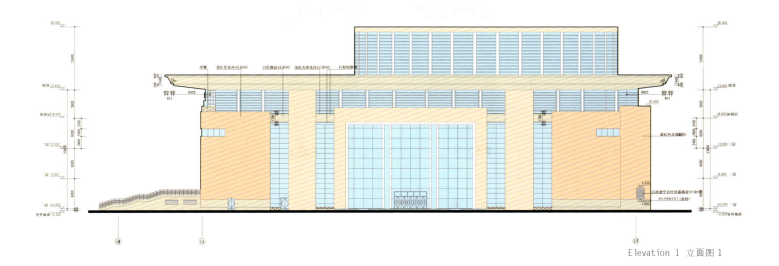

Elevation 1 立面图 1

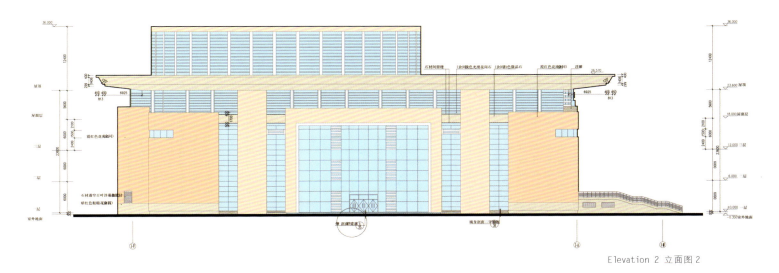

Elevation 2 立面图 2

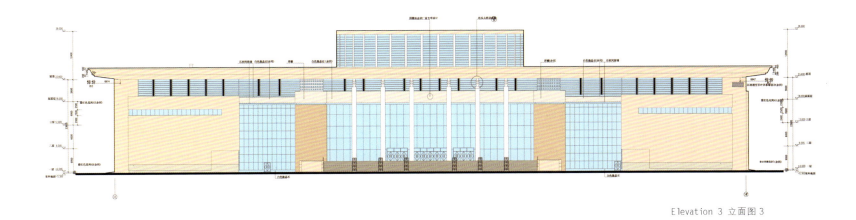

Elevation 3 立面图 3

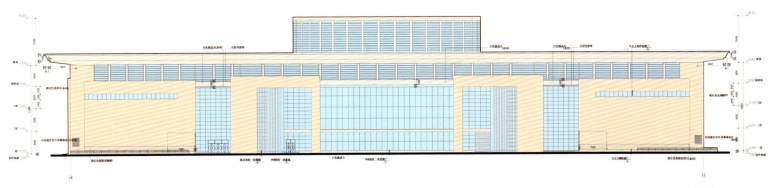

Elevation 4 立面图 4

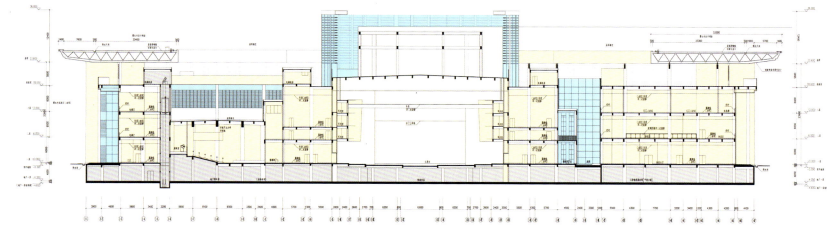

Section 1-1 剖面图 1-1

KEY WORDS 关键词

Three-dimensional Space 立体空间
Spirit of Place 场所感
Openness 开放性

Shenxianshu Community Service Center
神仙树社区服务中心

FEATURES 项目亮点

To avoid isolation between architecture and urban public space, a leisurely walkway is designed to run through the buildings. It not only provides more pubic spaces but also connects different functions together to show "openness" of public buildings.

设计打破建筑与城市公共空间分离的状态，在建筑中贯穿了一条供市民休闲的空间立体散步道，串联起建筑的多种功能，并充分体现了公共空间的开放性。

Location: Chengdu High-tech Zone, Sichuan, China
Architectural Design: China Southwest Architectural Design and Research Institute Corp. Ltd.
Land Area: approximately 9,000.3 m²
Total Floor Area: 13,834.38 m²
Building Density: 38.21%
Plot Ratio: 1.06
Green Coverage Ratio: 30.11%
Completion: 2011

项目地点：中国四川省成都市高新区
设计单位：中国建筑西南设计研究院有限公司
用地面积：约 9 000.3 m²
总建筑面积：1 3834.38 m²
建筑密度：38.21%
容积率：1.06
绿地率：30.11%
竣工时间：2011 年

Overview

Located in Zijing North Street of Chengdu High-Tech Zone, with two to five floors on ground and one floor underground, the service center boasts a total floor area of 13,834.38 m². It consists of the activity center and offices, supermarket, police station, health center and sports center of the community.

项目概况

项目位于成都市高新区紫荆北街，建筑地上 2~5 层，地下一层，总建筑面积为 1 3834.38 m²。建筑功能包括社区活动中心的相关功能及其办公区，以及社区超市、派出所、社区卫生院、社区体育健身中心。

Architectural Design

The architectural spaces are organized according to the functions, human behaviors and visual effects. Entries and exits as well as the pedestrian flows serve as the spatial nodes to defines the outdoor spaces and transition spaces. This kind of design will create multi-level spaces and enhance the spirit of the place. To avoid isolation between architecture and urban public space, a leisurely walkway is designed to run through the buildings. It not only provides more pubic spaces but also connects different functions together to show "openness" of public buildings.

建筑设计

建筑空间组织结合功能要求、人的行为特征及视觉感受，有机布局各功能出入口，合理组织交通流线，使每条流线及各出入口都成为空间景致的节点，使人在行进的过程中感受到室外空间与灰空间交替节奏的变换，这种使空间层次丰富多变的处理加强了社区的场所感。设计打破建筑与城市公共空间分离的状态，在建筑中贯穿了一条供市民休闲的空间立体散步道，即从本质上提供了使建筑成为公共性的空间的可能，这条"道"串联了建筑的多种功能，而且充分体现了公共空间的开放性。

Site Plan 总平面图

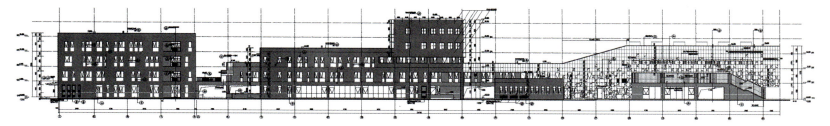
Elevation 1 立面图 1

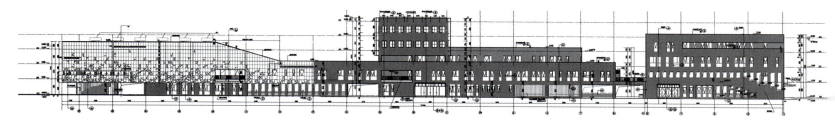
Elevation 2 立面图 2

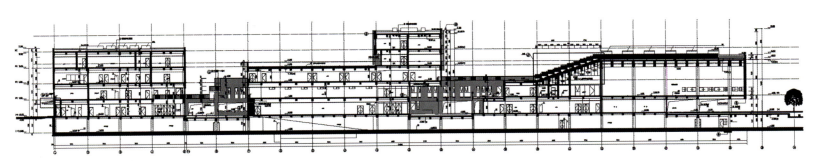

Sectional Drawing 剖面图

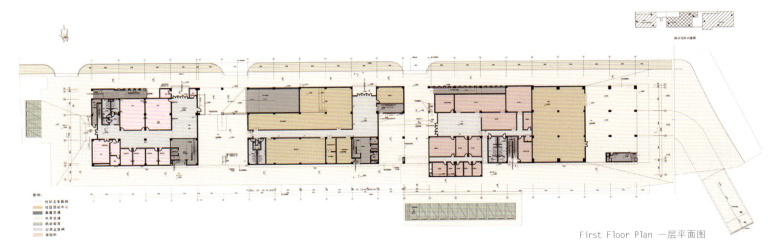

First Floor Plan 一层平面图

KEY WORDS 关键词

Chinese Classical Style 中式古典风格
Functional Layout 功能布局
Rich Facade 丰富立面

Water Park Reconstruction Project
水上公园改造工程

FEATURES 项目亮点

The reconstruction project covers the design of six monomer buildings that according to the function and location of each monomer, it adopts different classical architectural styles to make the overall building look well-proportioned with rich changes.

改造工程包括六个单体建筑的设计，每个单体根据各自的功能、坐落的位置，采用了不同的古典建筑风格，整体建筑错落有致，富于变化。

Location: Suzhou, Jiangsu, China
Architectural Design : Tianjin University Research Institute of Architecture Design
Total Floor Area: 11,207 m²
Office Building: 4,113 m²
Chengyin Building: 2,616 m²
Dongmen Loft: 1,313 m²
Shuibo Loft: 1,123 m²
Hubin Loft: 1,315 m²
Sandao Loft: 727 m²
Completion: 2010

项目地点：中国江苏省苏州市
设计单位：天津大学建筑设计研究院
总建筑面积：11 207 m²
办公楼：4 113 m²
澄赢楼：2 616 m²
东门轩：1 313 m²
水波轩：1 123 m²
湖滨轩：1 315 m²
三岛轩：727 m²
竣工时间：2010 年

Overview

The reconstruction project covers the design of six monomer buildings, i.e. Office Building, Chengyin Building, Dongmen Loft, Shuibo Loft, Hubin Loft and Sandao Loft. Dongmen Loft, Shuibo Loft, Hubin Loft and Sandao Loft are landscape architectures for visitors to rest and have a view. Office building is for the Bureau of Parks and Woods and Chengyin Building for food and beverage field.

项目概况

项目为水上公园改造工程，包括六个单体建筑的设计，六个单体分别为：办公楼、澄赢楼、东门轩、水波轩、湖滨轩、三岛轩。其中，东门轩、水波轩、湖滨轩、三岛轩为景观建筑，供游人休憩、观景使用；办公楼为园林局办公建筑，澄赢楼为餐饮建筑。

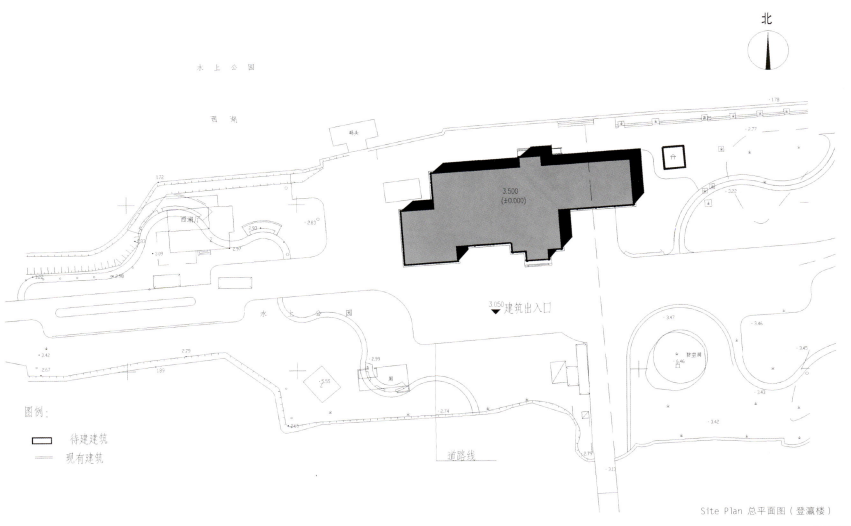

Site Plan 总平面图（登瀛楼）

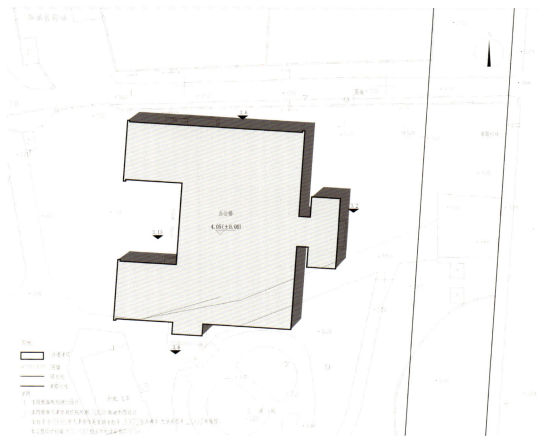

Design Concept

This project sites in the water park in order to coordinate with the big environment of the water park, and the series of these buildings all take Chinese classical architecture style. Based on the full study on the Chinese classical architecture, designers and experts coordinated to design such six monomers.

设计构思

项目建设地点在水上公园内，为了与水上公园的大环境相协调，此组建筑的风格均为中国古典建筑风格。在对中国古典建筑进行充分研究的基础上，设计人员与设计院的古建专家共同设计了此六个单体。

Office Building Site Plan 办公楼总平面图

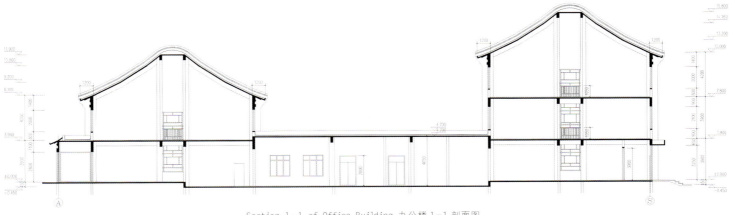

Section 1-1 of Office Building 办公楼 1-1 剖面图

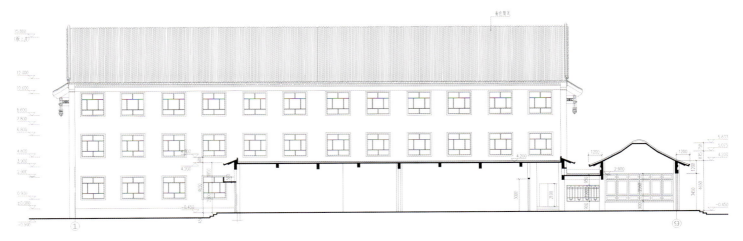

Section 2-2 of Office Building 办公楼 2-2 剖面图

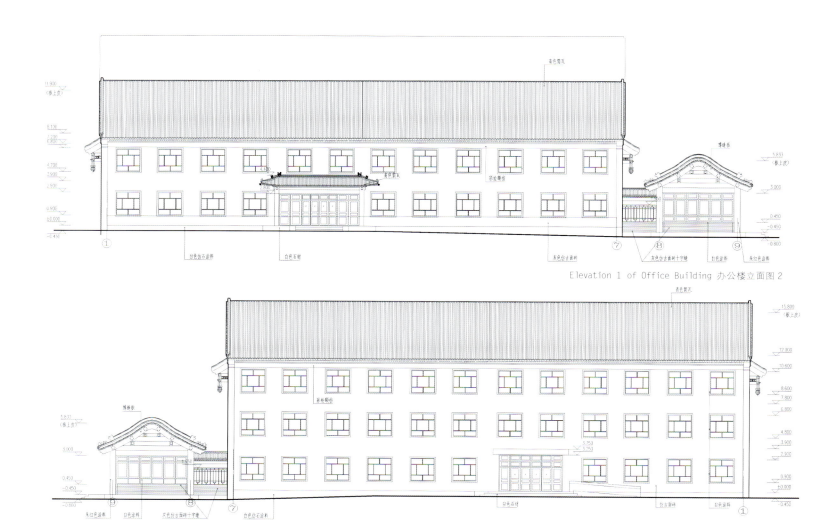

Elevation 1 of Office Building 办公楼立面图 2

Elevation 2 of Office Building 办公楼立面图 2

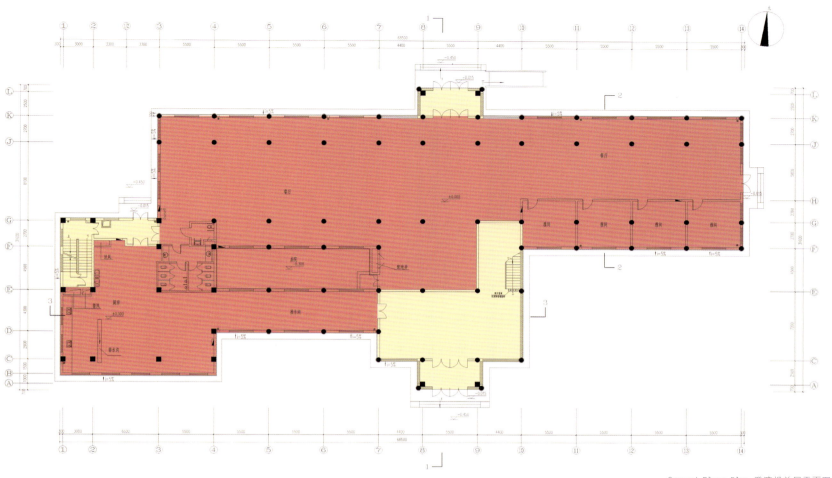

Ground Floor Plan 登瀛楼首层平面图

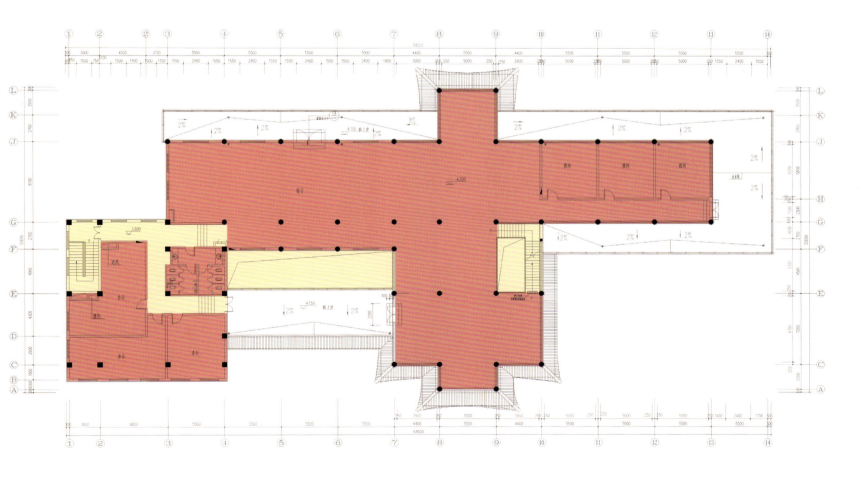

Second Floor Plan 登瀛楼二层平面图

Ground Floor Plan of Office Building 办公首层平面图

Architectural Design

According to the function and location of each monomer, each monomer building adopts different classical architectural styles. Office building uses flush gable roof classical architectural style, and this three-layer building has reasonable layout on the basis of meeting the functional requirements. Other buildings are gable and hip roof classical architecture, which take different changes according to different layouts. Chengyin Building is a three-layer catering building, of which the facade adopts pure classical architectural symbols and the plane combines with catering building characteristics with reasonable functional partitions. In order to activate the space and enrich the facades, it also sets terrace to make the overall building look well-proportioned with rich changes.

建筑设计

每个建筑单体根据各自的功能、坐落的位置，采用了不同的古典建筑风格。办公楼采用硬山古典建筑风格，为三层建筑，在满足功能需求的前提下，合理布局。其他建筑均为歇山古典建筑，根据平面布局不同，采用不同的层次变化。澄赢楼为三层餐饮建筑，立面采用纯古典建筑符号，平面结合餐饮建筑特点，合理划分功能分区。为活跃空间、丰富立面，设置露台的形式，整体建筑错落有致，富于变化。

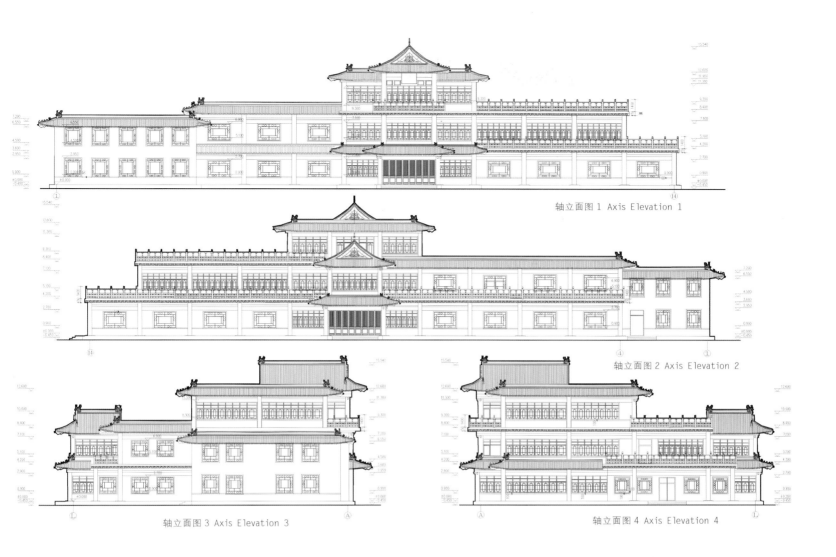

轴立面图 1 Axis Elevation 1

轴立面图 2 Axis Elevation 2

轴立面图 3 Axis Elevation 3

轴立面图 4 Axis Elevation 4

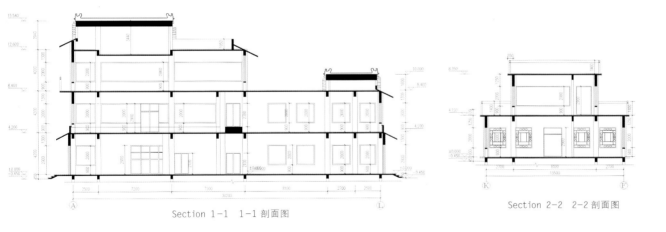

Section 1-1 1-1 剖面图

Section 2-2 2-2 剖面图

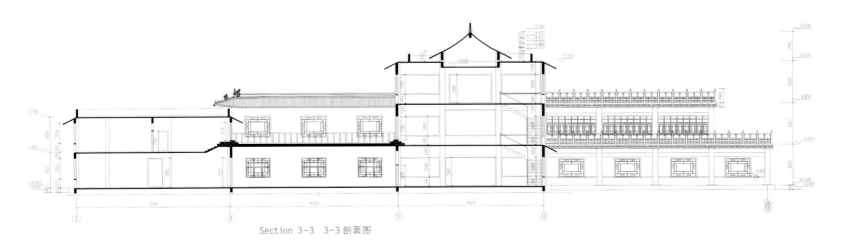

Section 3-3 3-3 剖面图

KEY WORDS 关键词

Roof Design 屋面设计
Lightsome Modeling 造型轻盈
Integrated Space 空间整合

Tianjin Meijiang Convention Center
天津梅江会展中心

FEATURES 项目亮点

The two roofing surfaces stretch gracefully like the wings and organically integrate the functional space together, which makes the exhibition center appear lightsome and looks as if fluttering and soaring high.

两组屋面舒展、优雅，犹如翅膀状，把功能空间有机地整合在一起，使会展中心外观轻盈，似振翅高飞。

Location: Jinnan District, Tianjin, China
Architectural Design: Tianjin Architecture Design Institute
Available Planning Area: 230,000 m²
Total Floor Area: 100,000 m²
Plot Ratio: 0.43
Completion: 2010

项目地点：中国天津市津南区
设计单位：天津市建筑设计院
规划可用地面积：230 000 m²
总建筑面积：100 000 m²
容积率：0.43
竣工时间：2010 年

Overview

Tianjin Meijiang Convention Center building has a total of 2 layers with a height of 36.3 m. The east side of the building is the main entrance for the audiences, 170 m away from the Friendship South Road, thus forming a 40,000 m² outdoor exhibition area. VIP entrance is set along the lake in the west with beautiful scenery, and the VIP car park is also arranged nearby. Cargo loading place in the middle part is both convenient and concealment. There are 800 car parking lots on the north and south sides for the guests, and parking space for non-motor vehicles and large trucks. The plane layout is developed from the function diagram of the large exhibition center. On both sides of the central hall, there organizes six exhibition hall in different sizes and sets up conference area at the end of the hall. The two roofing surfaces stretch gracefully like the wings and organically integrate the functional space together, which makes the exhibition center appear lightsome and looks as if fluttering and soaring high.

项目概况

天津梅江会展中心建筑共 2 层，高度为 36.3 m。建筑东侧为观众主入口，退友谊南路 170 m，形成 40 000 m² 的室外展场。贵宾入口在西侧沿湖设置，景色宜人，并就近设置贵宾停车场。展区中部的货物装卸区，既方便又隐蔽。南北两侧各安排 800 辆小汽车停车场，供来宾停车，并设有非机动车区和大型货车区。平面布局根据大型会展中心的功能简图发展而来。中央大厅两侧设置六个展厅，并在大厅尽端设置会议区。两组舒展、优雅，犹如翅膀状的屋面，把功能空间有机地整合在一起，使会展中心外观轻盈，似振翅高飞。

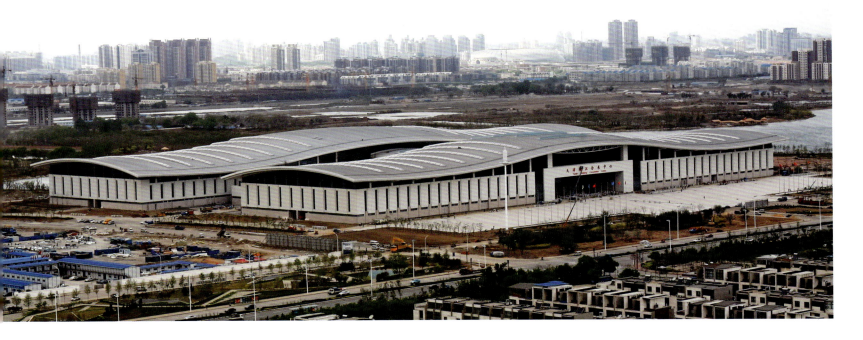

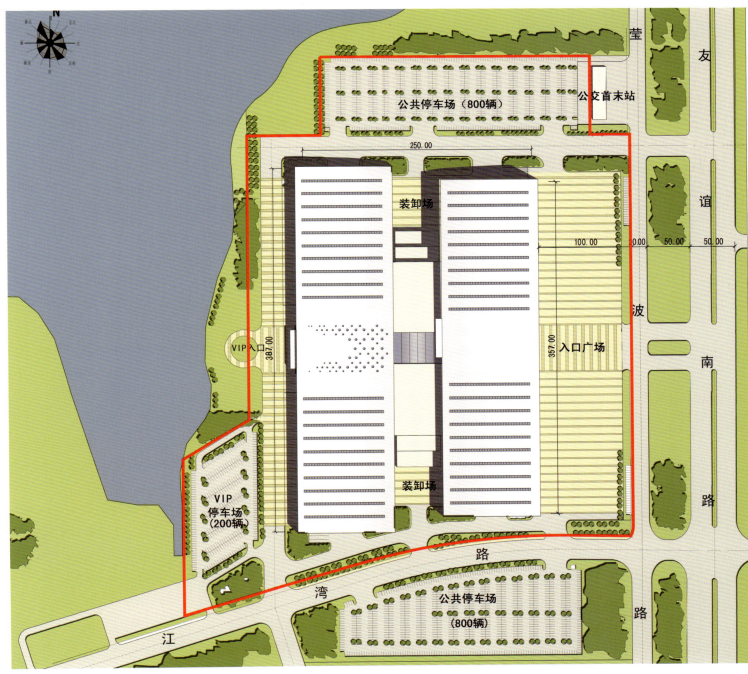

Site Plan 总平面图

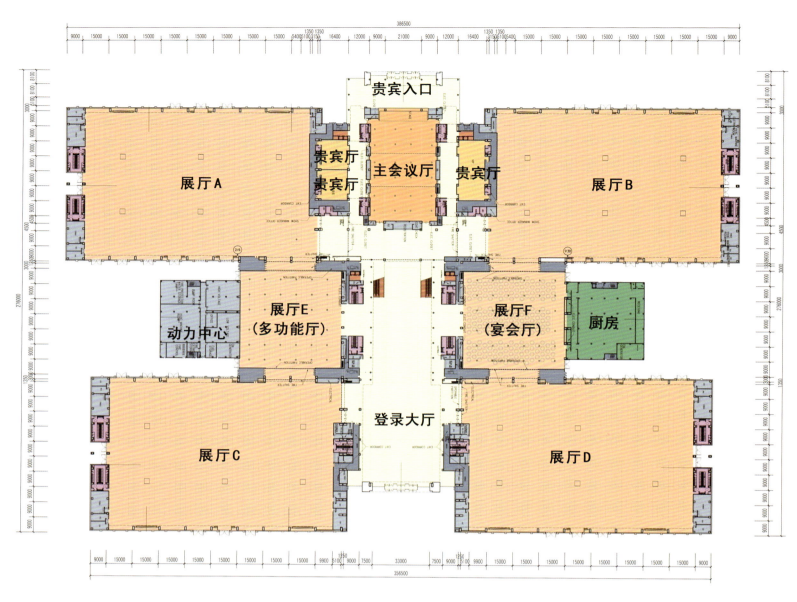

First Floor Plan 首层平面图

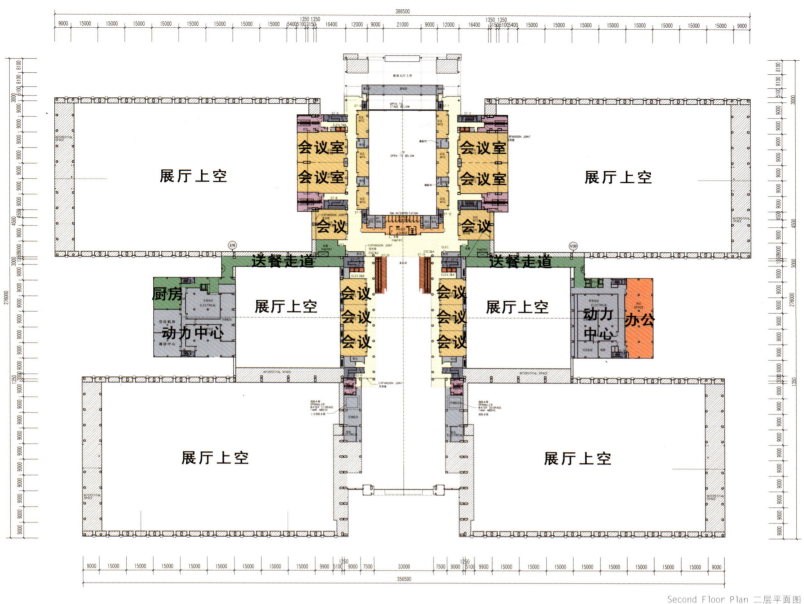

Second Floor Plan 二层平面图

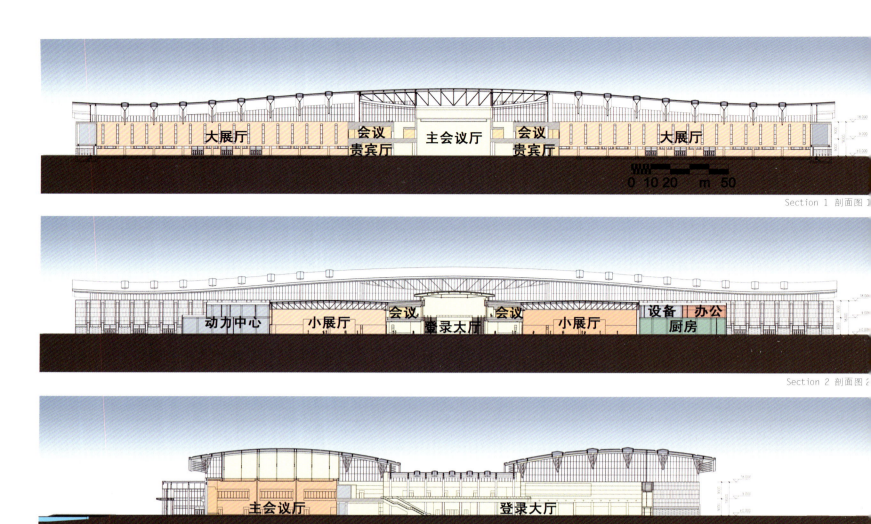

Section 1 剖面图1

Section 2 剖面图2

Section 3 剖面图3

登录大厅

大展厅

主会议厅

小展厅（宴会厅）

KEY WORDS 关键词

Lighting Atrium 采光中庭
Vertical Lines 竖向线条
Sense of Science and Technology 科技感

Xinjiang Science and Technology Museum Reconstruction and Extension Project
新疆科技馆改扩建工程

FEATURES 项目亮点

The main building continues to apply the uncapped intensive vertical lines up and down that stand for the endless science world, and at the upper left corner, there is a slant glazing, which indicates the window of the popular science. Exhibition hall of the annex building is designed into three-dimensional hexagonal mass, thus increasing its content of science and technology.

在主楼上继续延伸上下不封顶的密集竖线条，寓意"科学无止境"，并在左上角镶嵌一斜面玻璃窗，表示"科普之窗"。裙楼展厅设计成三维六角体量，增加科技含量。

Location: Uygur Autonomous Region, Xinjiang, China
Architectural Design: Xinjiang Institute of Architectural Design Research
Total Floor Area: 26,601.16 m²

项目地点：中国新疆维吾尔自治区
设计单位：新疆建筑设计研究院
总建筑面积：26 601.16 m²

Overview

Xinjiang Science and Technology Museum Reconstruction and Extension Project shows respect to the original pavilion's context. The main building continues to apply the uncapped intensive vertical lines up and down that stand for the endless science world, and at the upper left corner there is a slant glazing, which indicates the window of the popular science. Exhibition hall of the annex building is designed into three-dimensional hexagonal mass, thus increasing its content of science and technology. It has consistent mechanism with the original triangle and conveniently extends the hexagonal exhibition hall. The original small lighting atrium is expanded into four-layer large court, occupying 42% of the area of the overall museum.

项目概况

新疆科技馆改扩建工程尊重原馆的文脉。在主楼上继续延伸上下不封顶的密集竖线条，寓意"科学无止境"，并在左上角镶嵌一斜面玻璃窗，表示"科普之窗"。裙楼展厅设计成晶莹剔透的三维六角体量，增加科技含量。与原正三角形肌理一致，顺势扩展六边形展厅。将原有很小的展厅扩展成四层采光大中庭，占据全馆面积的42%。

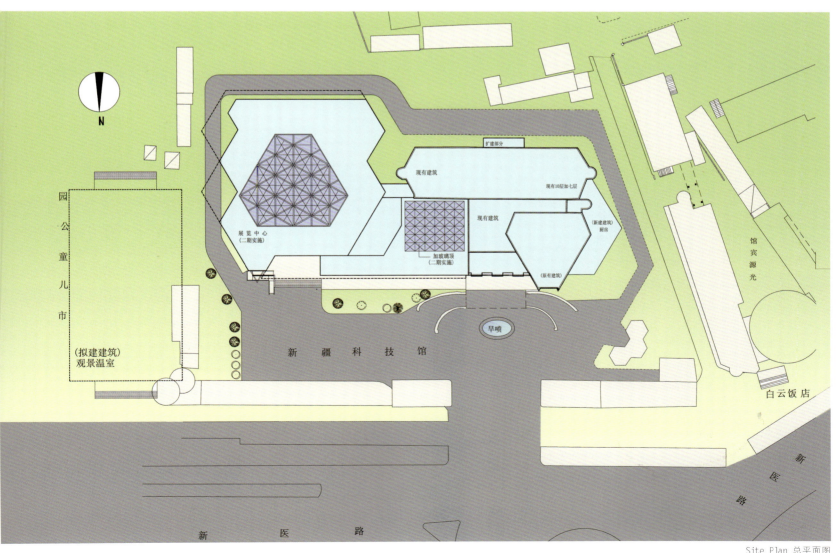

Site Plan 总平面图

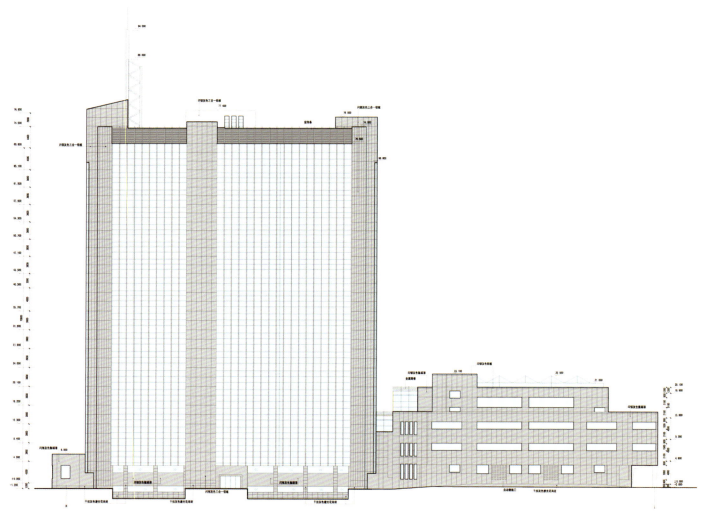

Axis Elevation 轴立面图

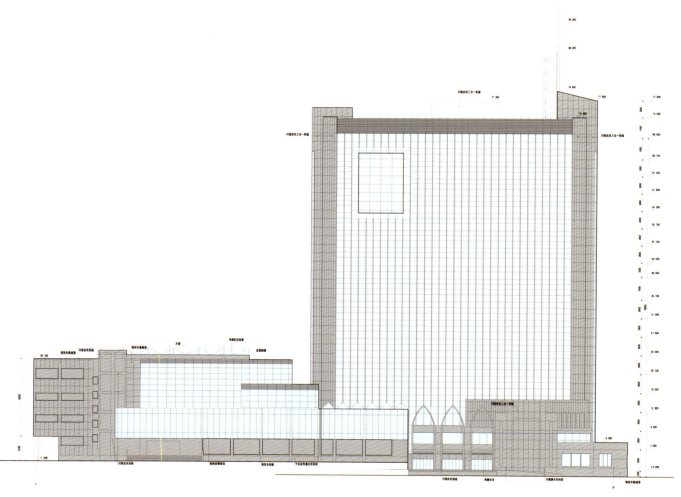

Axis Elevation 轴立面图

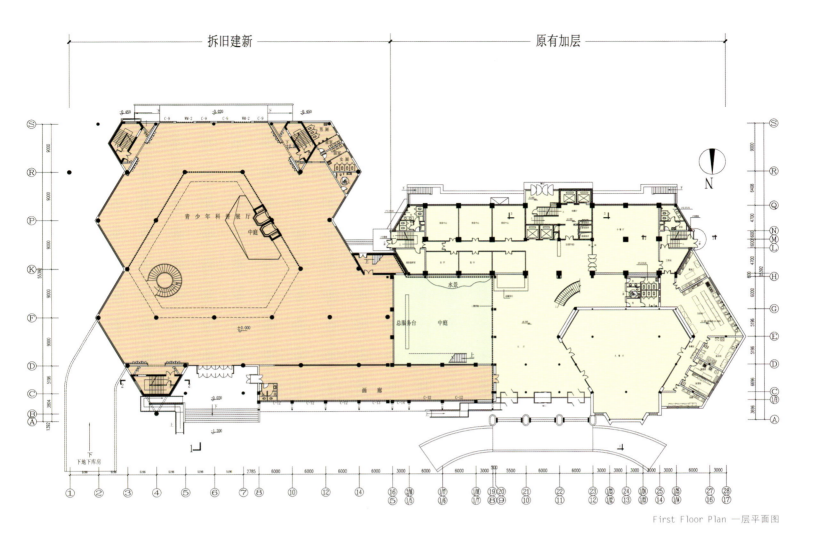

First Floor Plan 一层平面图

KEY WORDS 关键词

Building Color 建筑色彩
Glazed Tile 琉璃瓦
Facing Brick 面砖

Reconstruction and Extension Project of Guangdong Provincial Sun Yat-Sen Library
广东省立中山图书馆改扩建工程

FEATURES 项目亮点

According to the characteristics of the culture architecture, the project selects lively and elegant color to embody the Chinese traditional garden architecture features, and its light, simple but elegant exterior facades make the building naturally infuse into the surrounding environment, thus achieving the effect of harmony.

结合文化建筑的特点，选择明快和典雅的色彩体现中国传统庭院建筑特色，清淡、素雅的外墙立面风格，使建筑自然地融入周边环境之中，并达到和谐、统一的效果。

Location: Guangzhou, Guangdong, China
Architectural Design: Guangzhou Design Institute
Completion: 2010

项目地点：中国广东省广州市
设计单位：广州市设计院
竣工时间：2010 年

Overview

The project Phase I has a total land area of 59,611 m², a total construction area of 69,569 m², and includes Block A, B, C. Block A consists of revolution square (for the protection of cultural relics) and two-layer underground garage with building area of 19,273 m² and capability of holding 449 cars, Block B is Sun Yat-Sen Library (maintained in original state) with construction area of 31,612 m², one layer under the ground and ten layers above the ground, and the function is a library. Block C is the newly-built digital library with construction area of 20,253 m², four layers underground and seven layers above the ground, and the function is stack rooms.

项目概况

一期工程总用地为 59 611 m²，总建筑面积为 69 569 m²，分 A、B、C 三个区。A 区为革命广场（文物保护）及地下共二层车库，建筑面积 19 273 m²，可地下停车 449 辆；B 区为中山图书馆（原状维修），建筑面积 31 612 m²，地下一层、地上十层，功能为图书馆。C 区为新建数字化书库，建筑面积 20 253 m²，地下四层、地上七层，功能为书库。

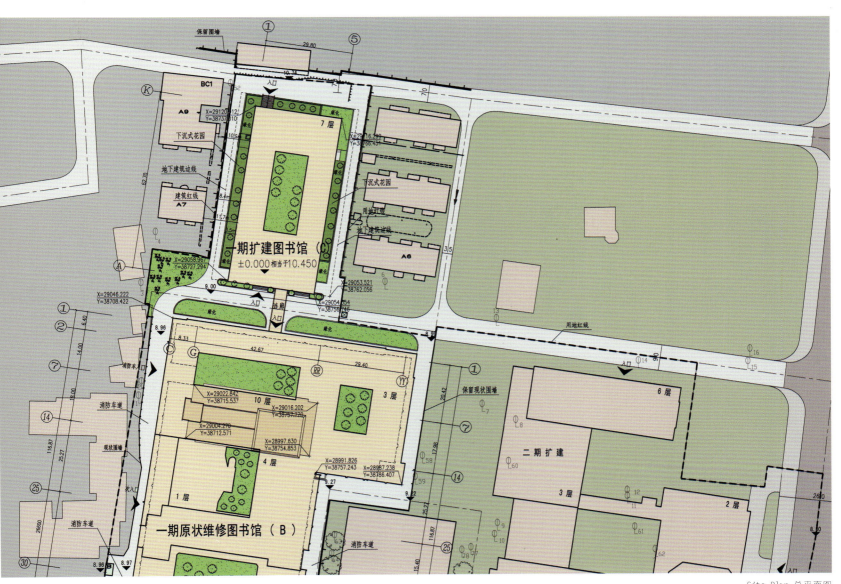

Site Plan 总平面图

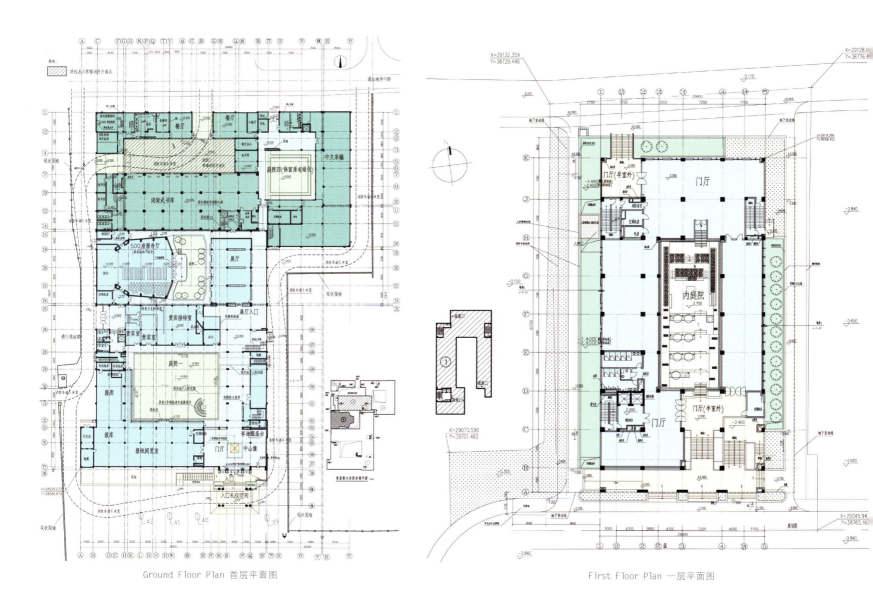

Ground Floor Plan 首层平面图

First Floor Plan 一层平面图

144

Architectural Design

According to the characteristics of the culture architecture, the color design and decorative materials of the exterior facade of Sun yat-sen Library select lively and elegant color to embody the Chinese traditional garden architecture features. The roof cover and cornice of Sun yat-sen Library adopt peacock blue glazed tile veneering, and the exterior facade uses light yellow facing bricks. The exterior facades of light, simple but elegant style can effectively reduce the building spatial volume amd make the building naturally infuse into the surrounding environment, thus achieving the effect of harmony

项目概况

中山图书馆建筑色彩设计和外立面装饰材料的选用，结合文化建筑的特点，选择明快和典雅的色彩体现中国传统庭院建筑特色。中山图书馆的屋面檐口采用孔雀蓝琉璃瓦贴面，建筑外墙面使用浅黄色面砖。清淡、素雅的外墙立面风格，可有效减小建筑的空间体量，使建筑自然地融入周边环境之中，并达到和谐、统一的效果。

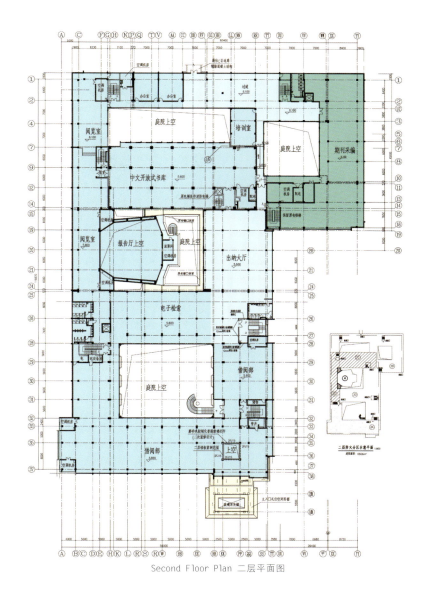

Second Floor Plan 二层平面图

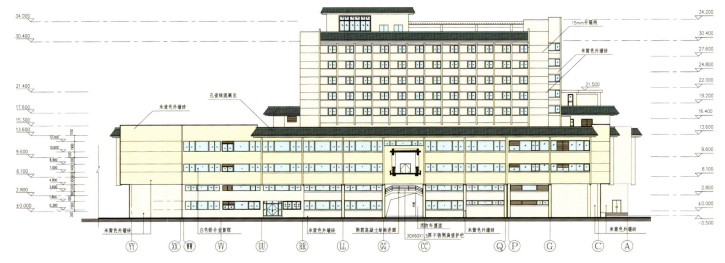
Elevation 1 立面图1

Elevation 2 立面图2

Elevation 3 立面图3

Elevation 4 立面图4

Section 1 剖面图 1 　　　　　　　　Section 2 剖面图 2

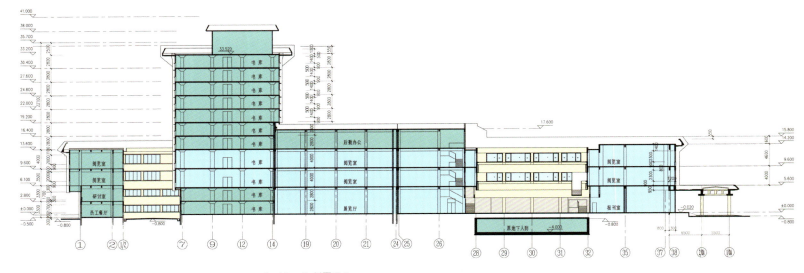

Section 3 剖面图 3

Elevation 5 立面图 5

Elevation 6 立面图 6

KEY WORDS 关键词

Spontaneous Expression 写意手法
Courtyard Space 院落空间
Cultural Ethos 文化气质

Ninghai Library
宁海县图书馆

FEATURES 项目亮点

In terms of architectural design, traditional Chinese technique "spontaneous expression" is introduced to create a series of changing courtyard space. White walls and black tiles are the keynote of this work. The exterior wall is dominated by soft and elegant greyish white paint and decorated partially with plain brick wall to reflect the sense of rhythm in details.

建筑造型上，以粉墙黛瓦为基调，外墙以柔性、淡雅的普通灰白涂料为主，局部装饰以清水砖墙，谱写细节上的节奏感。

Location: Ninghai, Ningbo, Zhejiang, China
Architectural Design: The Architectural Design & Research Institute of Zhejiang University Co., Ltd.
Land Area: 8,360 m²
Total Floor Area: 3,758 m²
Building Height: 9.9 m
Building Density: 27.3%
Plot Ratio: 0.45
Green Coverage Ratio: 32.1%
Completion: 2009

项目地点：中国浙江省宁波市宁海县
设计单位：浙江大学建筑设计研究院
用地面积：8 360 m²
总建筑面积：3 758 m²
建筑高度：9.9 m
建筑密度：27.3%
容积率：0.45
绿地率：32.1%
竣工时间：2009 年

Architectural Design

In terms of architectural design, traditional Chinese technique "spontaneous expression" is introduced to create a series of changing courtyard space. White walls and black tiles are the keynote of this work. The exterior wall is dominated by soft and elegant greyish white paint and decorated partially with plain brick wall to reflect the sense of rhythm in details. The simple greyish white is changing in accordance with the surrounding light, indicating the changing of the seasons. Neat and square volume interplays with well-aligned trees, and the changing light interweaves with the shadows, which narrows the distance between the building and environment. Designers aim at not only perfect harmony with Jiangnan courtyard, but cultural ethos of this building.

建筑设计

建筑单体设计上，借鉴了中国传统的写意造园的手法，构筑了一系列变化的院落空间。造型上，以粉墙黛瓦为基调，外墙以柔性、淡雅的普通灰白涂料为主，局部装饰以清水砖墙，谱写细节上的节奏感。简洁、质朴的灰白色跟随周围环境光线而变化，呈现出四时的变化，天光的投射，横斜的疏影，拉近了建筑与环境的距离。此举不仅是在风格意向与江南园林院落取得和谐一致的关系，而且使图书馆素净与刚正的外形与公园大环境错落有致的树木相互映衬，表现文化建筑所特有的纯净的书卷气质。

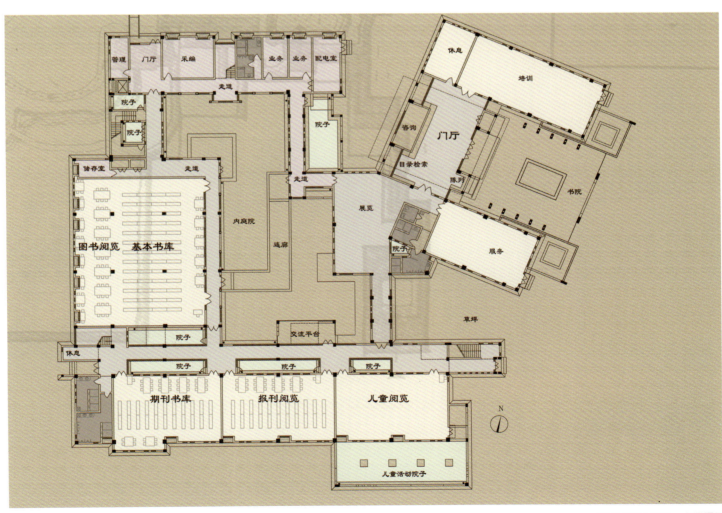

First Floor Plan 一层平面图

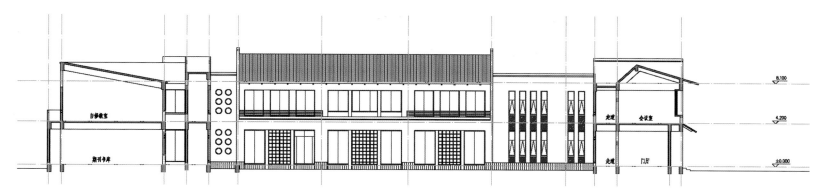

Section C-C C-C 剖面图

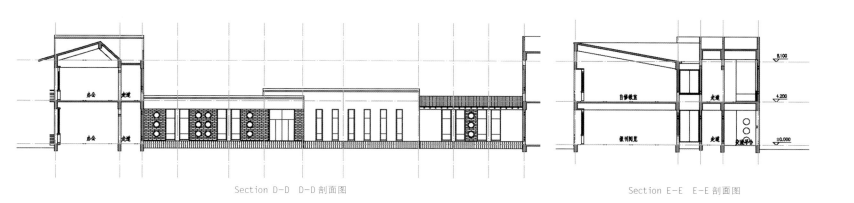

Section D-D D-D 剖面图   Section E-E E-E 剖面图

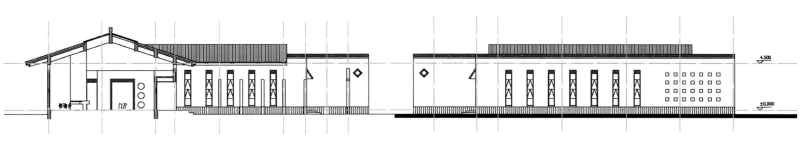

Section A-A A-A 剖面图   Elevation 1 立面图 1

Section B-B B-B 剖面图

Elevation 2 立面图 2

Elevation 3 立面图 3

Elevation 4 立面图 4

155

KEY WORDS 关键词

- **Jiangnan Charm** 江南神韵
- **Grass Roof** 植草屋顶
- **Construction Materials** 建筑材料

Hangzhou Cuisines Museum
杭帮菜博物馆

FEATURES 项目亮点

The main museum building adopts modern construction materials and technology to reproduce the spatial organization spirit and elements of traditional Hangzhou building, and uses construction technique of sloping roof, dry wall and full glass curtain wall to reflect the beautiful and refined charm of Hangzhou architecture.

博物馆主体建筑采用现代化的建筑材料和技术手段再现传统杭州建筑的空间组织精神和元素，通过坡屋顶、清水墙、全玻璃幕墙等建筑手法，体现杭州建筑的秀雅神韵。

Location: Hangzhou, Zhejiang, China
Architectural Design: China Architecture Design & Research Group
Design: 2010
Completion: 2012

项目地点：中国浙江省杭州市
设计单位：中国建筑设计研究院
设计时间：2010 年
竣工时间：2012 年

Overview

Hangzhou Cuisines Museum is located in Jiang Yang Fan Ecological Park of Hangzhou. It faces Qiantang River in the south, Lotus Peak in the north, Tiger Monastery in the west, and Yuhuang Mountain and Bagua Field in the east.

项目概况

中国杭帮菜博物馆，坐落在杭州市江洋畈原生态公园。南临钱塘江，北傍莲花峰，西连虎跑，东靠玉皇山、八卦田。

Architectural Design

Hangzhou Cuisines Museum as a cultural carrier is bound to reflect the regional characteristics of Hangzhou architecture. The regional characteristics not only refer the specific architectural elements of white wall, black tiles, veranda, and overhangs, but also lie in its overall spirit and temperament, i.e. the beautiful and refined charm of Hangzhou. The designers hope to use modern architectural materials and technical means to transfer the spirit and temperament, and reproduce the spatial organization spirit of traditional Hangzhou building, thus making it a modern one in Hangzhou.

The architecture design of Hangzhou Cuisines Museum takes in the form of a continuous sloping roof to flexibly resolve the large-volume roof, so as to form a unified and attractive architecture form and space changes full of Jiangnan architectural charm. The overlapping and continuous roof echo with the surrounding mountain formation; at the same time, the green grass roof makes the building truly integrat into the overall environment.

建筑设计

杭帮菜博物馆作为一个文化载体，必然要体现杭州建筑的地域性特色。这一地域性特色不仅仅在于白墙黑瓦游廊挑檐等这些具体的建筑构成要素，更在于其整体所彰显出来的精神气质，具体就杭州而言，就是其"秀"、"雅"的神韵。设计师希望用现代化的建筑材料和技术手段彰显建筑的这种精神气质，再现传统杭州建筑的空间组织精神，使之成为现代化的杭州建筑。

杭帮菜博物馆建筑设计以连续坡屋顶的形式将大体量的建筑屋顶进行巧妙的分解，形成统一而又富韵味的建筑形体和空间变化，极富江南建筑神韵。连续而又层层重叠的屋顶和周边山体形成呼应之势，同时，绿色植草屋顶也使建筑真正地融入整体环境之中。

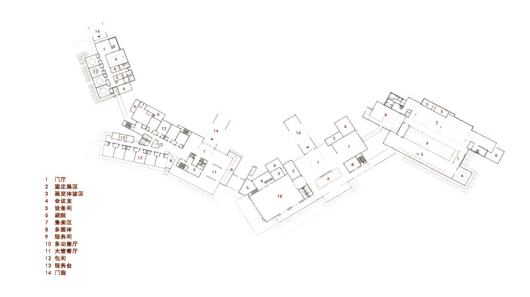

1 门厅
2 固定展区
3 展览体验区
4 会议室
5 设备间
6 庭院
7 售卖区
8 多媒体
9 服务间
10 多功能厅
11 大堂餐厅
12 包间
13 服务台
14 门廊

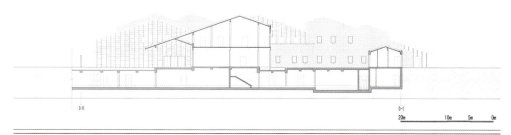

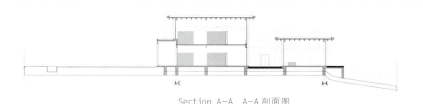

Section A-A A-A 剖面图

KEY WORDS 关键词

Artificial Stone Modeling 仿奇石造型
Streamline 流线状
Facade Texture 立面肌理

Liuzhou Wonder Stone Museum
柳州奇石馆

FEATURES 项目亮点

The construction conception infuses the nature beauty perceived from the wonder stones into the architectural form of the stone museum, which makes the building look like the moving mountain or stone, and adds more unique charm of the stone into the entire building.

建筑构思将从观赏奇石领悟到的大自然美感融入奇石馆的建筑形体中，建筑形体犹如奇峰或奇石变化，为整个展览馆增添了独特的石之韵味。

Location: Liuzhou, Guangxi Zhuang Autonomous Region, China
Architectural Design: Tianjin University Research Institute of Architectural Design & Urban Planning
Land Area: 125,000 m²
Site Area: 6,865.14 m²
Total Floor Area: 12,392.3 m²
Building Height: 22.13 m
Plot Ratio: 0.2
Green Coverage Ratio: 76%
Completion: 2012

项目地点：中国广西壮族自治区柳州市
设计单位：天津大学建筑设计规划研究院
用地面积：125 000 m²
占地面积：6 865.14 m²
总建筑面积：12 392.3 m²
建筑高度：22.13 m
容积率：0.2
绿地率：76%
竣工时间：2012 年

Architectural Design

The architectural form of Wonder Stone Museum has solid and flexibility, Yin and Yang as an, which looks is like the moving mountain or stone, and the grids on the facade are based on the patterns and skin texture of the earth, mountains, and rocks, and adds more unique charm of the stone into the entire building. For the facade design, it uses computer parametric approach to constitute the grids with four module units, and these units are all created by the same pattern. The four basic units, through random rotation, mirroring and combination, subtly reproduce four kinds of methods, i.e. protruding, concaving, supporting and removing, thus maintaining the unity of the facade while creating more abundant texture changes. The front facade and back elevation have made distinct design. Grids of the front facade are stacked horizontally to show the smooth lines; and the ones of the back facade are vertically piled, embodying the tallness and straightness of the rock. The construction conception infuses the nature beauty perceived from the wonder stones into the architectural form of the stone museum, the overall building is not an ending but a flowing and changing process.

建筑设计

奇石馆建筑形体融山水的坚柔于一体，犹如奇峰或奇石变化，立面上的栅格则脱胎于大地、山体，以及岩石表面的纹路和肌理，为整个展览馆增添了独特的石之韵味。立面设计运用计算机参数化方式，使栅格由四种模数制式的单元构成，单元全部由同一图案生成。四种基本单元随机旋转、镜像并组合的过程中，又巧妙地衍生出凸出、凹进、衬底、剔除四种手法，从而在保持立面统一性的同时，又形成更为丰富的肌理变化。建筑正立面和背立面做了区别设计。正面栅格横向叠砌，展示纹之流畅；背立面则竖向垒叠，表现岩石之挺拔。建筑构思将从观赏奇石领悟到的大自然美感融入奇石馆的建筑形体中，建筑整体不是一个结果而是一个流动、变化的过程。

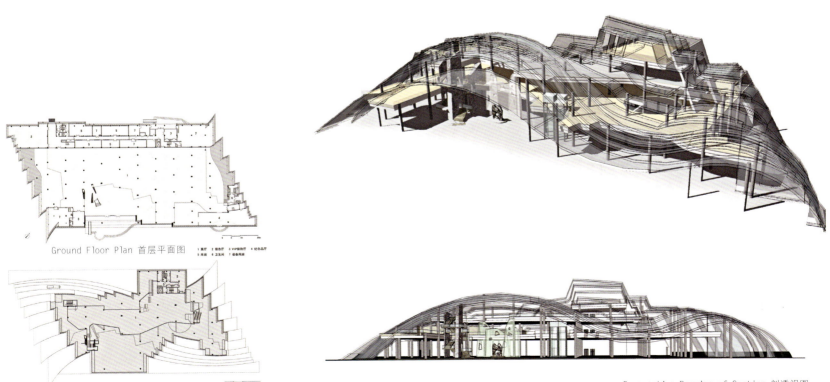

Ground Floor Plan 首层平面图

Third Floor Plan 三层平面图

Perspective Drawing of Section 剖透视图

Site Plan 总平面图

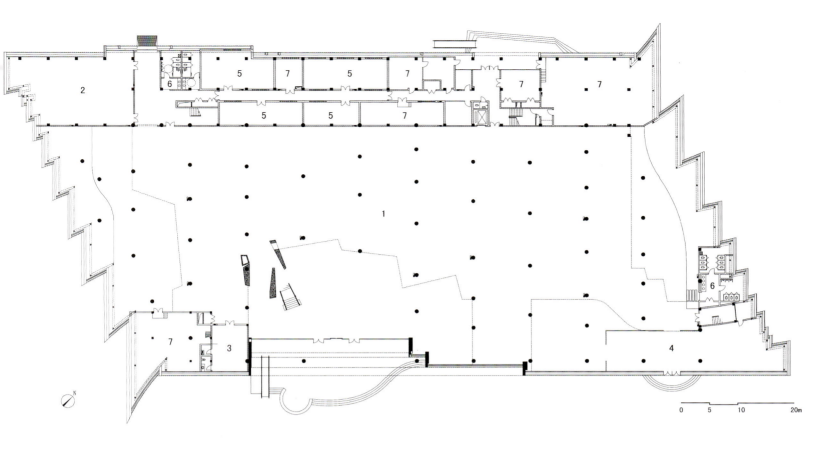

1 展厅　2 报告厅　3 VIP接待厅　4 纪念品厅
5 库房　6 卫生间　7 设备用房

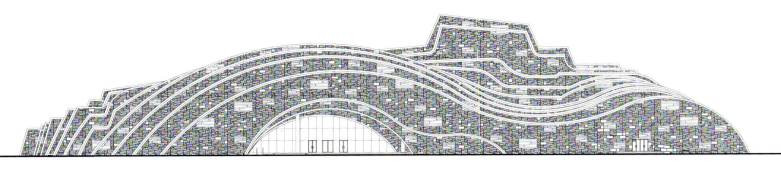

Elevation 1 立面图1

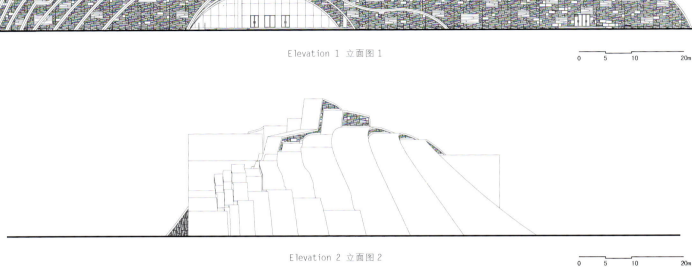

Elevation 2 立面图2

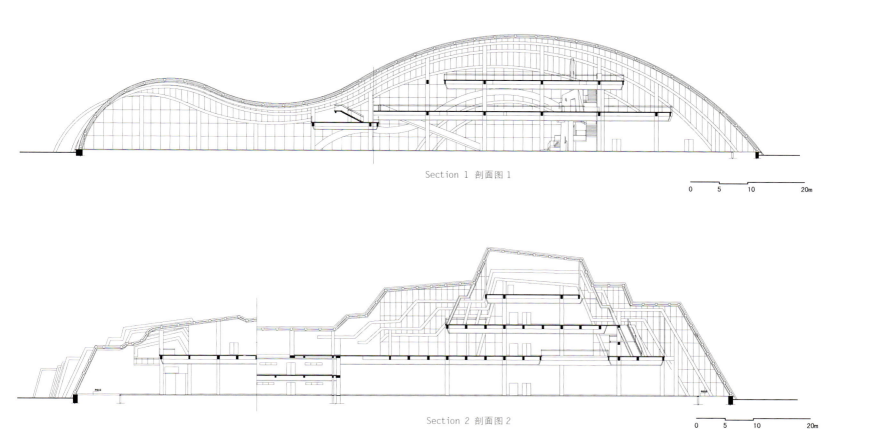

Section 1 剖面图 1

Section 2 剖面图 2

ized_test
Sports Building
体育建筑

- Large-span Structure 大跨度结构
- Varied Shapes 造型各异
- Spectacular Volume 气势恢宏
- Highly Intelligent 高智能化

KEY WORDS 关键词

Two-tone Spiral Flow 双色螺旋流动造型

Blue and White Modeling 蓝白相间

Rectangular Body Mass 矩形体量

2010 Guangzhou Asian Games Swimming and Diving Hall of Provincial Venues
2010年广州亚运会省属场馆游泳跳水馆

FEATURES 项目亮点

For swimming functional requirements, the architectural modeling is dealt to show the gradient formation level and the overall form of gentle undulating. By boldly using the blue and white colors, the spaced colors form an interspersed dynamic styling, which makes the rectangular massing very dynamic.

结合游泳馆的功能要求,建筑造型被处理成层级渐变、和缓起伏的总体形态,设计师大胆运用蓝、白两种颜色,间隔布置,形成穿插对比的动态造型,使矩形的体量展现出丰富的动感。

Location: Guangzhou, Guangdong, China
Architectural Design: Architectural Design and Research Institute, South China University of Technology
Land Area: 99,978 m²
Total Floor Area: approximately 33,331 m²
Plot Ratio: 0.71

Interior Design

By interspersed modeling design forming levels of interior space, the design meets the different needs of interior space for swimming, diving and training. By elevation the training pool one level in the building, the design can on the one hand obtain a compact interior space height, reducing the use of energy; on the other hand, its lower part of the room pool deck has been fully utilized, while reducing the basement area and lowing the cost.

内部设计

设计方案通过造型的穿插,形成室内空间的高低之差,从而满足了游泳、跳水、训练对于室内空间的不同需求。在建筑内部把训练池抬高一层,一方面可以形成紧凑的室内空间高度,减少使用能耗;另一方面池岸下部的房间得到了充分利用,同时减少了地下室面积,降低造价。

项目地点:中国广东省广州市
设计单位:华南理工大学建筑设计研究院
用地面积:99 978 m²
总建筑面积:约33 331 m²
容积率:0.71

Site Plan 总平面图

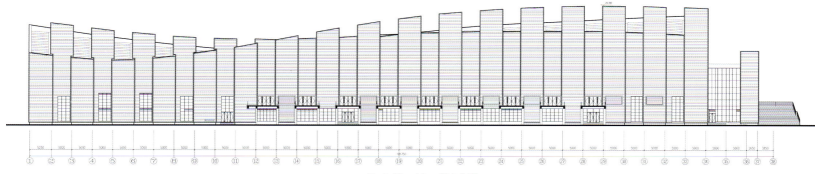
West Elevation 西立面图

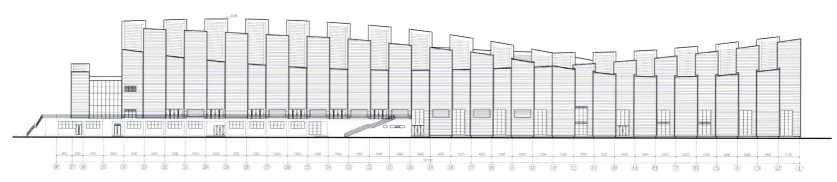
East Elevation 东立面图

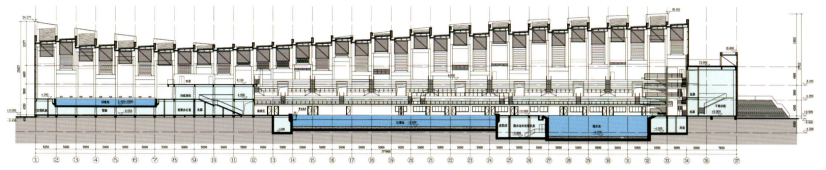

Section 1 剖面图 1

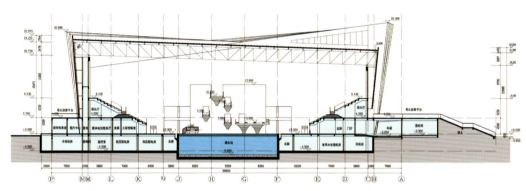

Section 2 剖面图 2

Ground Floor Plan 首层平面图

Architectural Design

For swimming functional requirements, the architectural modeling is dealt to show the gradient formation level and the overall form of gentle undulating. By boldly using the blue and white colors, the spaced colors form an interspersed dynamic styling, which makes the rectangular massing very dynamic. By using the spiral flow modeling and white and blue intertwines, the design both cleverly metaphor Guangzhou "Yunshan beads of water," the city geographical features and a continuation of the main stadium "streamers" curve. Meanwhile, interspersed flow modeling combined with building orientation, the design well meets the height of the interior space of the building, lighting and ventilation needs, energy efficiency and rational arrangement of equipment pipeline requirements.

建筑设计

结合游泳馆的功能要求，建筑造型被处理成层级渐变、和缓起伏的总体形态，设计师大胆运用蓝、白两种颜色，间隔布置，形成穿插对比的动态造型，使矩形的体量展现出丰富的动感。主体造型采用双色螺旋流动造型，主体建筑蓝白相间，既巧妙地隐喻了广州"云山珠水"的城市地理特征，又是对主体育场"飘带"曲线的延续。同时通过相互穿插流动造型，结合建筑朝向，很好地满足了建筑内部空间高度、采光通风、建筑节能以及合理布置设备管道的需求。

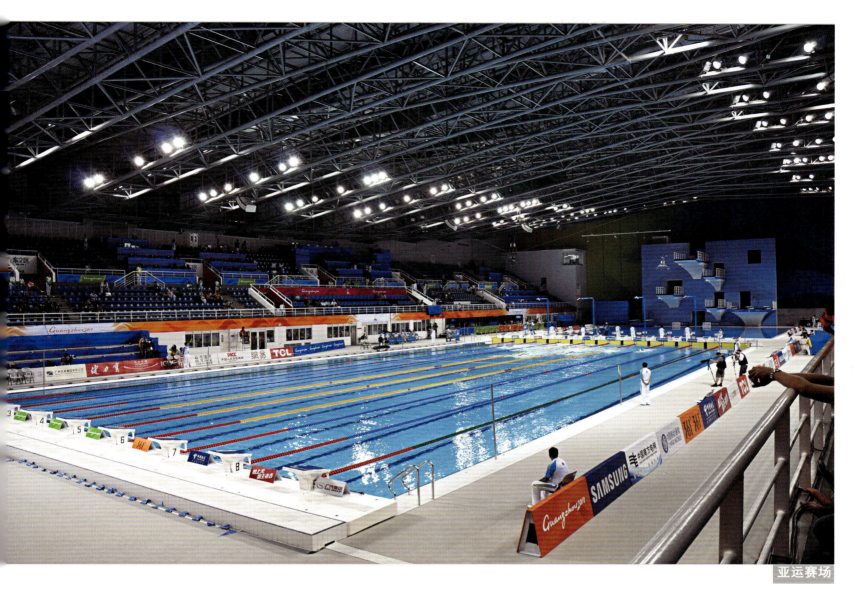

亚运赛场

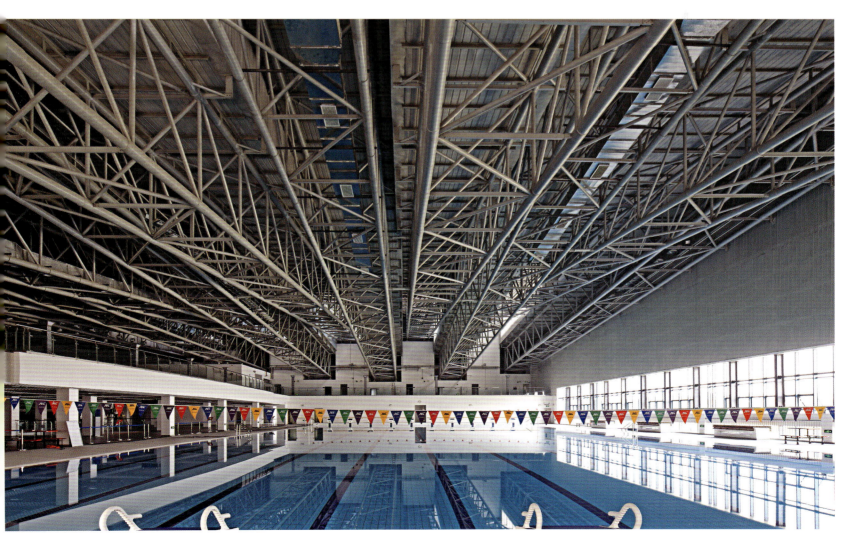

KEY WORDS 关键词

Roofing Inflection 反曲屋面
Unique Shape 造型独特
Smooth Lines 线条流畅

South China Technology University Stadium, Guangzhou University City
广州大学城华南理工大学体育馆

FEATURES 项目亮点

Consisting of four twisted shell which made of a combination of anti-curved roof in its design, the stadium shows a smooth shape and vitality.

在造型设计上，体育馆由四个扭壳组合而成的反曲屋面呈现出流畅的造型和蓬勃向上的生命力。

Location: Guangzhou, Guangdong, China
Architectural Design: Architectural Design and Research Institute, South China University of Technology
Total Land Area: 19,779 m^2
Total Floor Area: 12,783 m^2
Underground Construction Area: 8,490 m^2
Ground Floor Area: 4,293 m^2
Total Building Height: 32.05 m

项目地点：中国广东省广州市
设计单位：华南理工大学建筑设计研究院
总用地面积：19 779 m^2
总建筑面积：12 783 m^2
地下建筑面积：8 490 m^2
地上建筑面积：4 293 m^2
建筑总高度：32.05 m

Architectural Design

Consisting of four twisted shell which made of a combination of anti-curved roof in its design, the stadium shows a smooth shape and vitality.

At the highest point of the roof, four intersected support structure set up an export component with the wind pulling exhaust fan. It serves both internal ventilation of the stadium and also becomes a highlight of the shape of the roof highlighting the stadium memorial image.

Colonnade, inflection roofing and architectural elements such as corner warped represent important architectural image of the national character of the old campus; the design of the new stadium has been re-interpreted by the language of modern architecture with the same purpose.

建筑设计

在造型设计上，体育馆由四个扭壳组合而成的反曲屋面呈现出流畅的造型和蓬勃向上的生命力。

在四个支撑构架相交的屋面最高点处，设置了一个配备排风机的拔风构件，既是体育馆内部通风换气的出口，也成为屋面造型的亮点，突出体育馆的纪念性形象。

代表华南理工大学老校区重要形象的传统柱廊、反曲屋面以及屋角起翘等建筑元素，在新体育馆设计中被现代建筑语言重新演绎，具有异曲同工之妙。

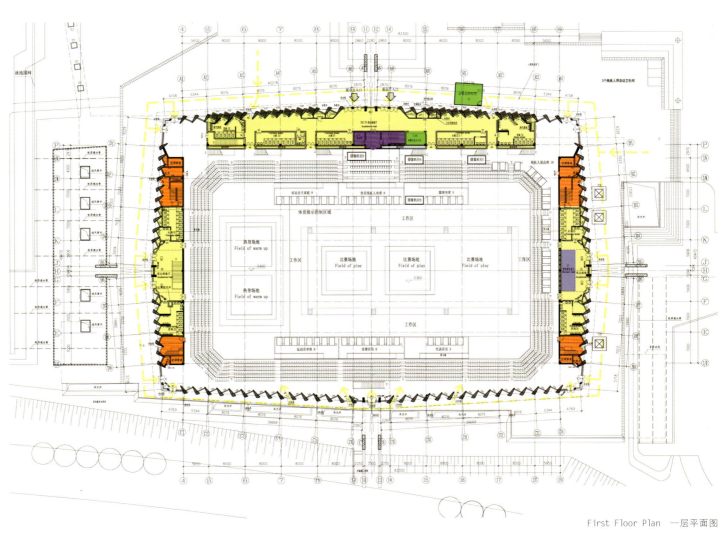

First Floor Plan 一层平面图

183

Site Plan 总平面图

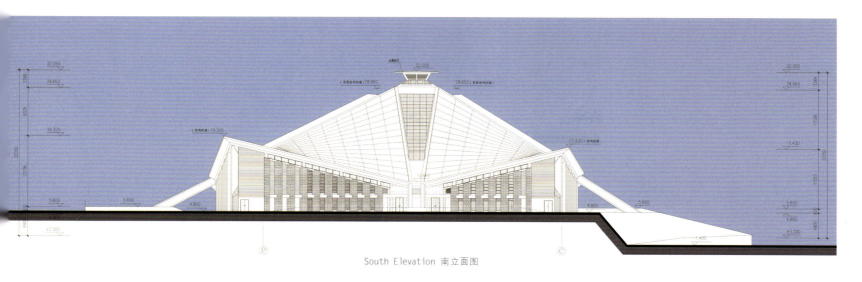
South Elevation 南立面图

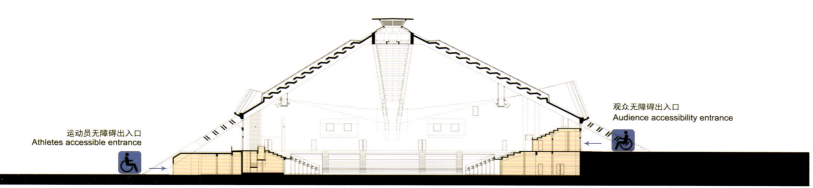
Cross Section 横剖面图

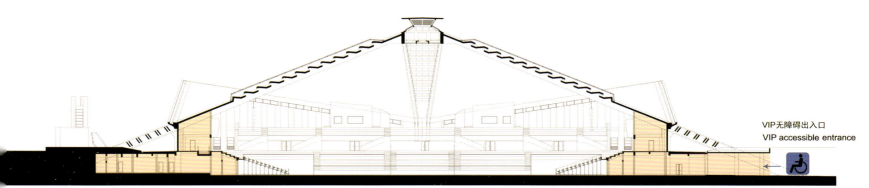
Longitudinal Section 纵剖面图

KEY WORDS
关键词

Diamond Shape 钻石造型

Open Roof 可开启屋盖

V-type Structure V 形结构

National Tennis Hall
国家网球馆

FEATURES
项目亮点

National Tennis Hall consists of multiple separate building units around together, resembling a bright diamond shape, very elegant, therefore also known as "Diamond Stadium".

国家网球馆由多组独立的建筑单元围绕而成，外形颇似一颗璀璨的钻石，十分典雅，因此，也被称为"钻石球场"。

Location: Chaoyang District, Beijing, China
Architectural Design: China Architecture Design & Research Institute
Land Area: 170,000 m²
Total Floor Area: 51,199 m²
Building Height: 46 m
Plot Ratio: 0.44

Overview

National Tennis Hall is mainly for professional tennis tournament with the needs of multi-purpose. It will become the main stadium for China open special game and will be Beijing region's largest tennis stadium with advanced international level. After the game it will serve for tennis competitions, training-based multi-functional sports venue and cultural center, and will provide sports, recreation, fitness, commerce and other integrated services.

项目地点：中国北京市朝阳区
设计单位：中国建筑设计研究院
用地面积：170 000 m²
总建筑面积：51 199 m²
建筑高度：46 m
容积率：0.44

项目概况

国家网球馆是以专业网球比赛为主，兼顾多功能需求的体育场馆。建成后将成为承办"中国网球公开赛"的专用比赛主场馆，为北京地区最大且具有国际先进水平的网球比赛场馆。赛后作为以网球比赛、训练为主的多功能体育、文化活动中心场馆，并可提供运动、休闲、健身和商业等综合性服务。

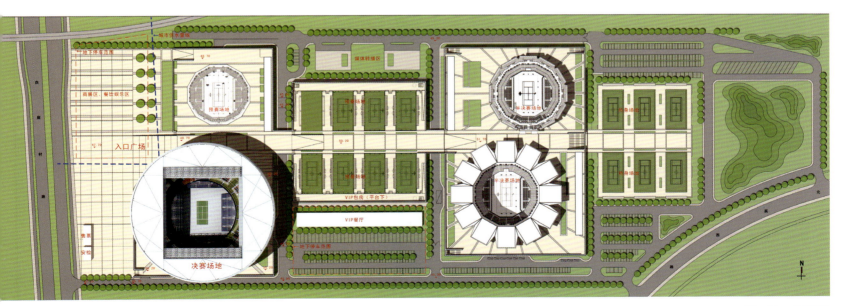

Site Plan 总平面图

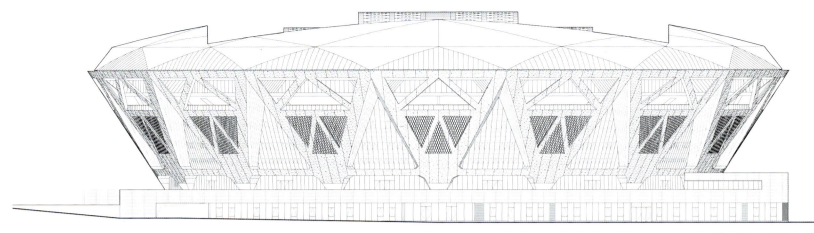

Elevation 立面图

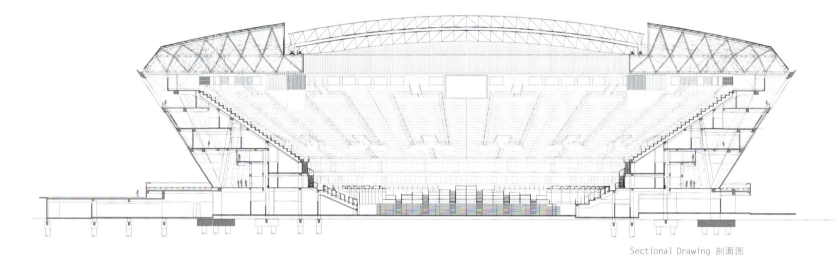

Sectional Drawing 剖面图

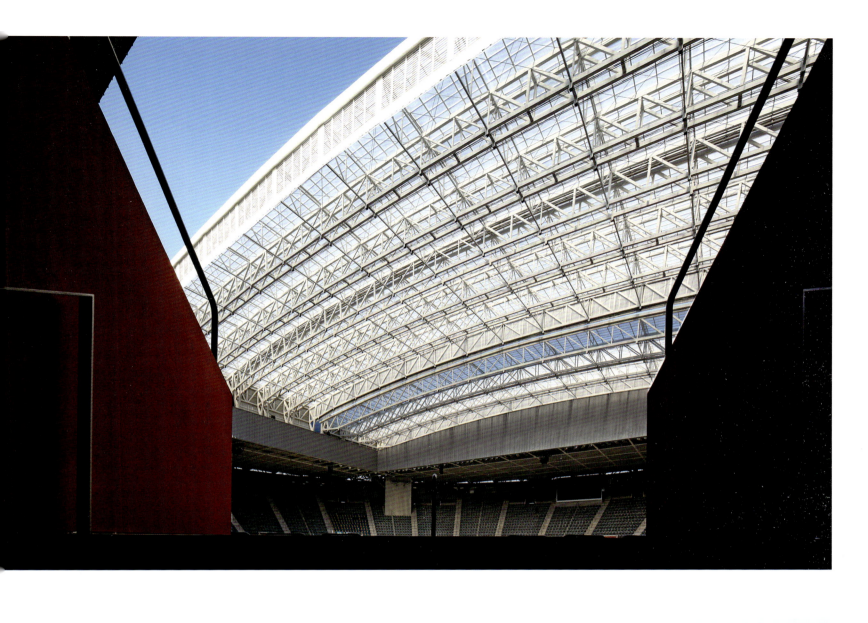

First Floor Plan 一层平面图

Architectural Design

National Tennis Hall consists of multiple separate building unit around together, resembling a bright diamond shape, very elegant therefore, also known as "Diamond Stadium". In order to meet th requirements of network events and multi-functional use require ments, the building has been designed to open the roof. Stadium circular layout design is reasonable, with no blind spots on t grandstand seats benefitting from reasonable roof open way an short opening time; a V-shape structural support and building ou look, making it both a uniform Lotus style and a strong personality

建筑设计

国家网球馆由多组独立的建筑单元围绕而成，外形颇似一颗璀璨的钻石，十分典雅，因此，也被称为"钻石球场"。为了满足中网的赛事要求及多功能使用要求，建筑采用了可开启屋盖的设计体育场圆形的平面布局设计合理，看台席位的视点无死角；屋顶开启方式合理，开启时间短；V形的结构支撑和建筑造型，使之与莲花球场既风格统一，又具有鲜明的个性。

KEY WORDS 关键词	Graceful Appearance 外观大气
	Fully Functional 功能齐全
	Modern Art 现代艺术

Luoyang New District Sports Center Stadium
洛阳新区体育中心体育场

FEATURES 项目亮点

The completion of the project makes it the central plains region's largest, most versatile and most advanced facilities of comprehensive sports center.

项目的建成使洛阳新区体育中心成为中原地区规模最大、功能最全、设施最先进的综合性体育中心。

Location: Luoyang, Henan, China
Architectural Design: Architectural Design and Research Institute of Tsinghua University
Land Area: 46,230 m²
Total Floor Area: 45,220 m²
Plot Ratio: 0.24

Overview

Located in an artificial lake on the west side of Luoyang New District Sports Center, west to university road, the stadium is one of the main venues of Luoyang New District Sports Center Project Phase II, it meets the national track and field, soccer competition and various large-scale community activities.

项目地点：中国河南省洛阳市
设计单位：清华大学建筑设计研究院有限公司
占地面积：46 230 m²
总建筑面积：45 220 m²
容积率：0.24

项目概况

洛阳新区体育中心体育场建设地点位于洛阳新区体育中心人工湖西侧，西邻大学路，是洛阳新区体育中心二期工程的主要场馆之一，可举办国家级田径、足球项目比赛及各类大型社会活动。

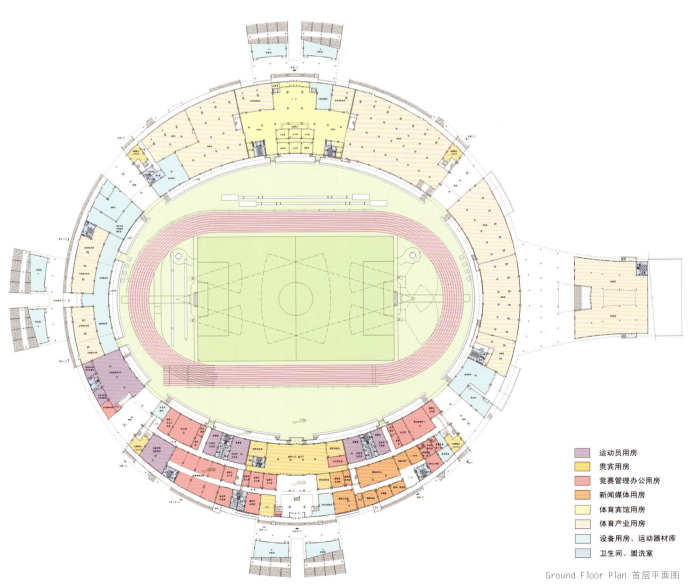

- 运动员用房
- 贵宾用房
- 竞赛管理办公用房
- 新闻媒体用房
- 体育宾馆用房
- 体育产业用房
- 设备用房、运动器材库
- 卫生间、盥洗室

Ground Floor Plan 首层平面图

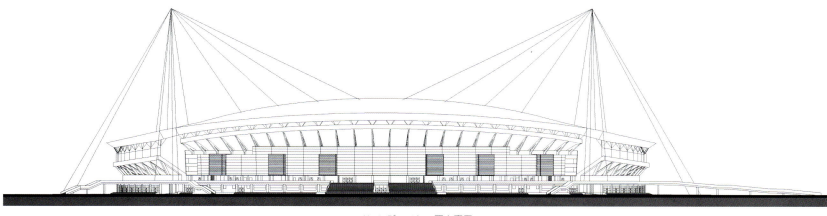

West Elevation 西立面图

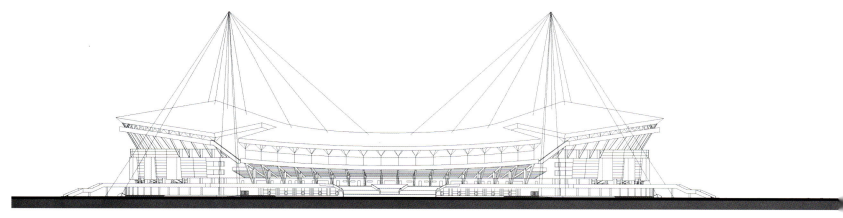

South Elevation 南立面图

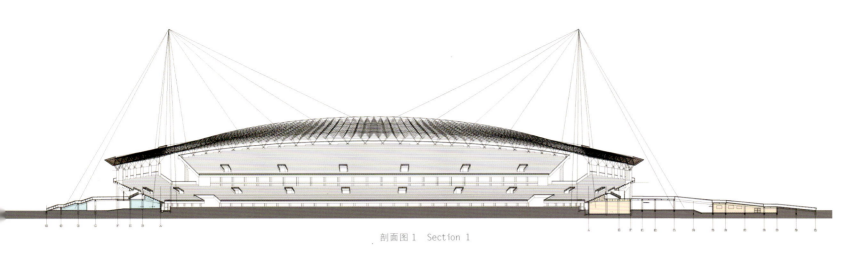

剖面图 1　Section 1

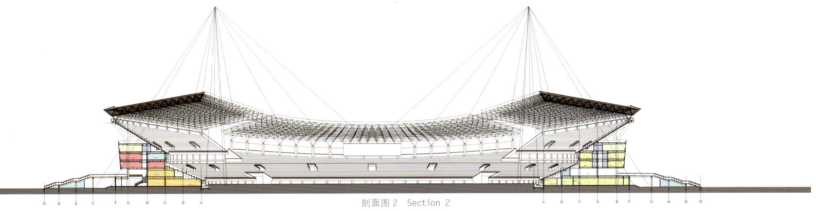

剖面图 2　Section 2

Layout

Corner venue divided passages and auxiliary channels into four relatively independent partitions at each side. Where, the VIP area, event organizers, office, athletes, referees lounge, media houses, etc. in the west, sport hotel with reception capacity of about 400 people in the eastern district; equipment rooms, warehouses and other sports equipment in the north and south zones. The audience is divided into upper and lower levels, where the site is located on both sides of the upper bleachers between the two houses of VIP boxes and control equipment. The podium at the lower west stands central location. Audience via stairs on the second floor from four directions into the auditorium, the upper audience through a dedicated stairs to the seating area without interfere the first floor functional houses staffs. The completion of the project makes it the central plains region's largest, most versatile and most advanced facilities of comprehensive sports center.

规划布局

四角场地通道将看台和辅助用房分为东、西、南、北四个相对独立的分区，其中：西区为贵宾区，包括赛事组委会、办公、运动员、裁判员休息室、媒体用房等；东区为体育宾馆，接待能力约400人左右；南北区为设备用房、体育器材仓库等。观众席分为上、下两层，其中，上层看台位于场地东西两侧，两层看台间为贵宾包厢及设备控制用房，主席台位于西看台下层中央位置。观众经由四个方向的大台阶由二层进入观众席，上层观众则通过专用楼梯到达座席区，与首层的功能用房人流互不干扰。该项目的建成使洛阳新区体育中心成为中原地区规模最大、功能最全、设施最先进的综合性体育中心。

KEY WORDS 关键词

- Curved Element 曲面单元
- Circular String Dome 环形张弦穹顶
- Steel Structure 钢结构

Nansha Gymnasium
南沙体育馆

FEATURES 项目亮点

Nine curved elements formed the shell of Gymnasium, elements stacked piece by piece, and divided into south and north groups, with the center of the game hall as their center, spreading spiral radically which divides the single building volume into two spaces and closely integrates them via dynamic way.

组成体育馆外壳的九个曲面单元，单元间片片层叠，并分为南北两组，以比赛大厅圆心为中心呈螺旋放射状展开，将单一的建筑体量一分为二，并以一种富有动感的方式将两者紧密联系。

Location: Guangzhou, Guangdong, China
Architectural Design: Architectural Design Institute of South China University of Technology
Total Floor Area: 30,236 m²
Plot Ratio: 0.18

Overview

Nansha Gymnasium is used as the venue of Wushu and Kabaddi matches, with 8,000 seats (6,000 fixed seats and 2,000 active seats). With steel and reinforced concrete structure, the major steel structure of the game hall applies advanced circular string dome, and its main span reaches 98 m.

项目地点：中国广东省广州市
设计单位：华南理工大学建筑设计研究院
总建筑面积：30 236 m²
容积率：0.18

项目概况

南沙体育馆建成后作为2010年广州亚运会武术及卡巴迪比赛馆，总坐席数为8 000座（其中固定坐席6 000，活动坐席2 000）。南沙体育馆采用了钢筋混凝土结构及钢结构，比赛大厅主体钢结构部分采用了先进的环形张弦穹顶技术，主跨度达到了98 m。

Site Plan 总平面图

Planning and Layout

The plane layout of Nansha Gymnasium borrows the classical approach of Yoyogi Gymnasium—the rotunda extends with tangential direction, develops into two spacious lobbies with clear sense of the direction, and on this basis, warm-up training venue is arranged on the side of the stadium, it combines with the platform on the second floor, develops into semi-open spaces for activity and rest that suits subtropical climate of Guangzhou. The classic plane form is extremely similar to Taiji diagram, and his coincidence also gives designers the opportunity to take advantage of the concept of Wushu. Nine curved elements formed the shell of Gymnasium, elements stacked piece by piece, and divided into south and north groups, with the center of the game hall as their center, spreading spiral radically, which divides the single building volume into two spaces and closely integrates them via dynamic way, the shape of the building is approximate to Taiji diagram, which is a metaphor for supreme state of Chinese Wushu—"Yin and Yang unite, harmony between man and nature".

平面布局

南沙体育馆的平面布局，借鉴了代代木体育馆的经典平面布局处理手法——圆形的比赛大厅，沿切线方向外延，形成两个宽敞而方向感明确的休息厅；在此基础上，在场馆的一侧设置了热身训练馆，结合二层平台，形成了适合广州亚热带气候特点的半开敞休息、活动空间。这种经典的平面形态与太极图构成极为相似，这种巧合也给了设计师巧妙利用"武术"概念的机会。组成体育馆外壳的九个曲面单元，单元间片片层叠，并分为南北两组，以比赛大厅圆心为中心呈螺旋放射状展开，将单一的建筑体量一分为二，并以一种富有动感的方式将两者紧密联系，运用近似太极图的构成方式，隐喻中华武术的至高境界——"阴阳俱合，天人合一"。

205

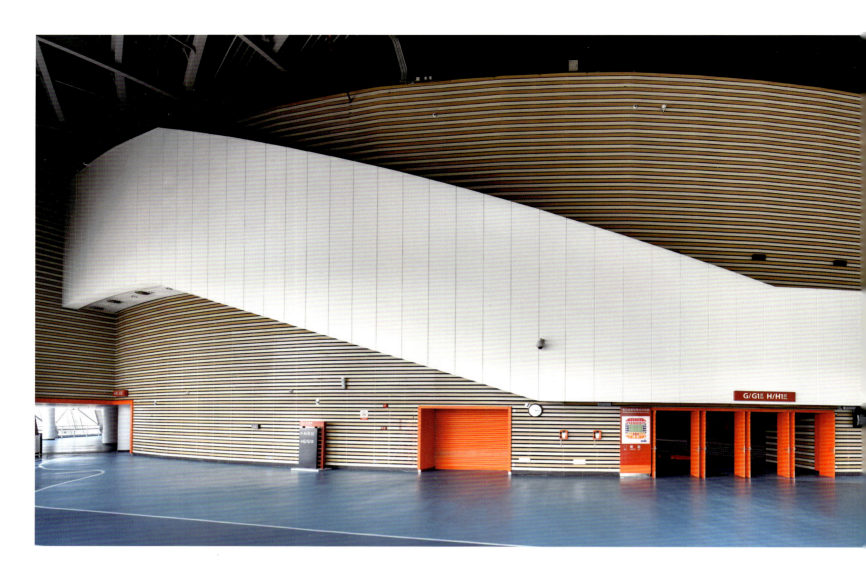

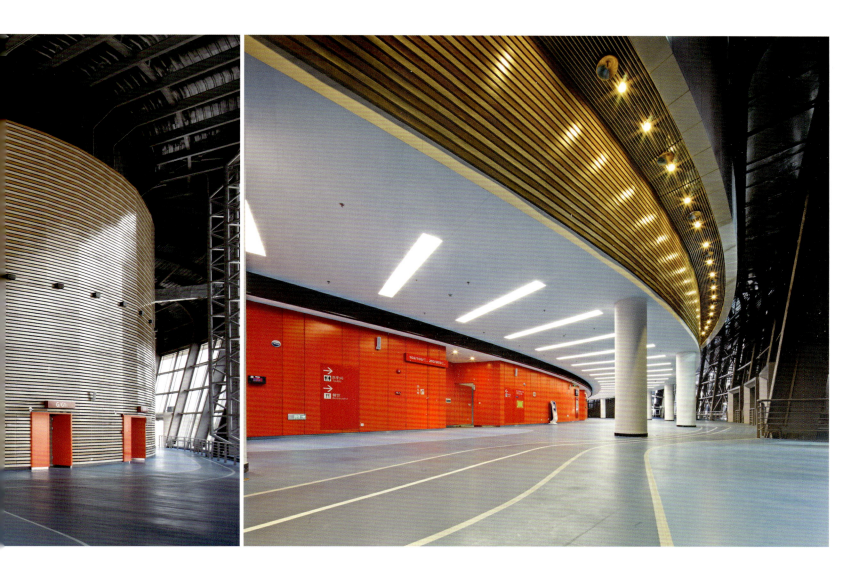
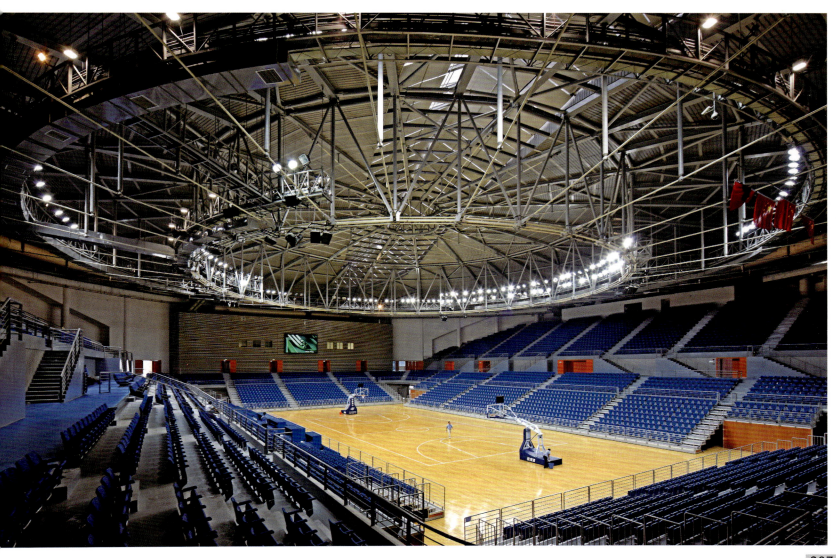

KEY WORDS 关键词

Curve Modeling 曲线造型
Modern Materials 现代材料
Concise Exterior 外观简洁

Putian Sports Center
莆田市体育中心

FEATURES 项目亮点

The building shape fully reflects the organic combination of internal function, spatial relationship and structure system. In the use of architectural language, curve, curved surface, aluminum plate, glass and other modern materials are used in this project, making the building full of extremely contemporary feeling.

在形体上，建筑造型的内部功能、空间关系和结构系统三者有机地融为一体。在建筑语言运用上，曲线、曲面及铝板和玻璃等现代材料的综合运用，使建筑极具现代感。

Location: Putian, Fujian, China
Architectural Design: Architectural Design & Research Institute of Tongji University (Group) Co, Ltd
Total Floor Area: 47,422 m^2
Land Area: 100,100 m^2
Plot Ratio: 0.474

Overview

Putian Sports Center contains 6,000 gymnasiums and 800 natatoriums. The gymnasiums are 5 levels aboveground, and the building height is 32.40 m, the natatoriums are 1 level underground, and 2 levels aboveground, and the building height is 16.70 m.

项目地点：中国福建省莆田市
设计单位：同济大学建筑设计研究院（集团）有限公司
总建筑面积：47 422 m^2
用地面积：100 100 m^2
容积率：0.474

项目概况

体育中心包含有6 000座体育馆和800座游泳馆。体育馆地上5层，建筑高度为32.40 m；游泳馆地下1层，地上2层，建筑高度为16.70 m。

Architecture Modeling Design

The building shape fully reflects the organic combination of internal function, spatial relationship and structure system. In the use of architectural language, curve, curved surface, aluminum plate, glass and other modern materials are used in this project, making the building full of extremely contemporary feeling. The multi-purpose sports-stadium with simplified shape; the free window in the side of the building not only ensures the needs of internal lighting, but also makes the whole building more brisk and lively, embodying the design concept of "ocean melody".

建筑造型设计

在形体上，建筑造型的内部功能、空间关系和结构系统三者有机地融为一体。在建筑语言运用上，曲线、曲面及铝板和玻璃等现代材料的综合运用，使建筑极具现代感。综合体育馆造型简洁，侧面自由的开窗既保证了内部采光的需求，也使得整个建筑显得轻快、活泼，体现了"海韵"的设计理念。

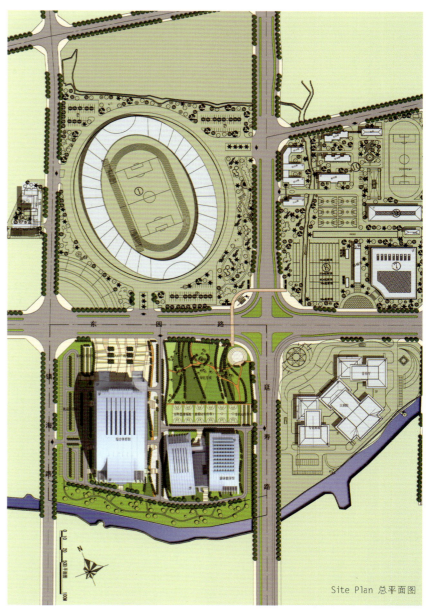

Site Plan 总平面图

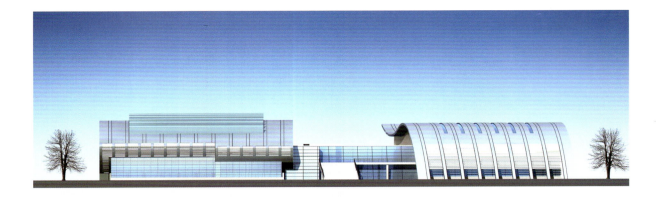

North Elevation of the Natatorium 游泳馆北立面图

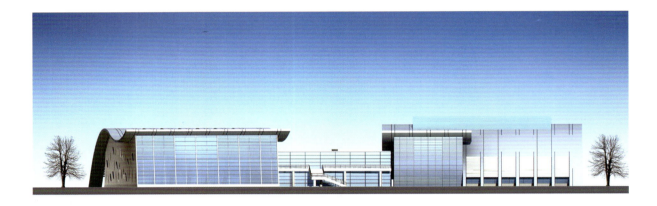

South Elevation of the Natatorium 游泳馆南立面图

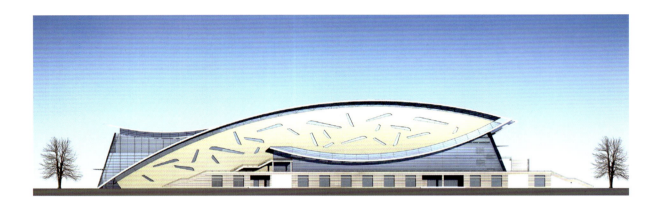

East Elevation of Gymnasium 体育馆东立面图

North Elevation of Gymnasium 体育馆北立面图

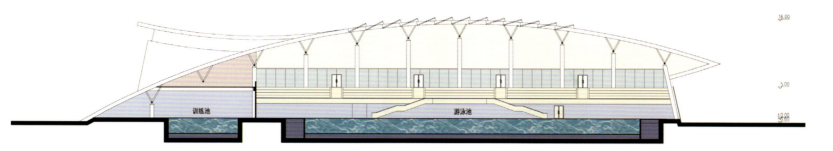

Section 1-1   1-1 剖面图

Section 2-2   2-2 剖面图

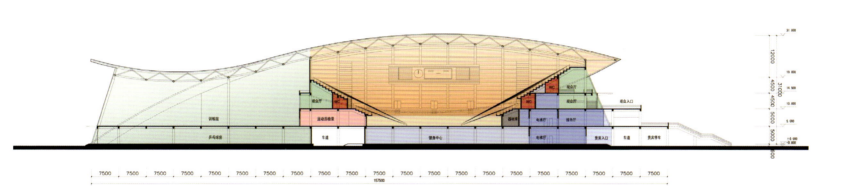

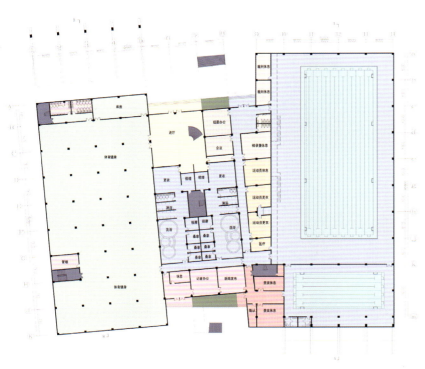
First Floor Plan of the Natatorium 游泳馆一层平面图

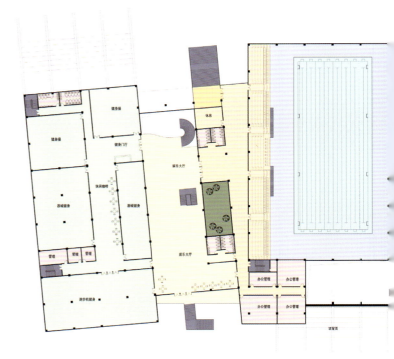
Second Floor Plan of the Natatorium 游泳馆二层平面图

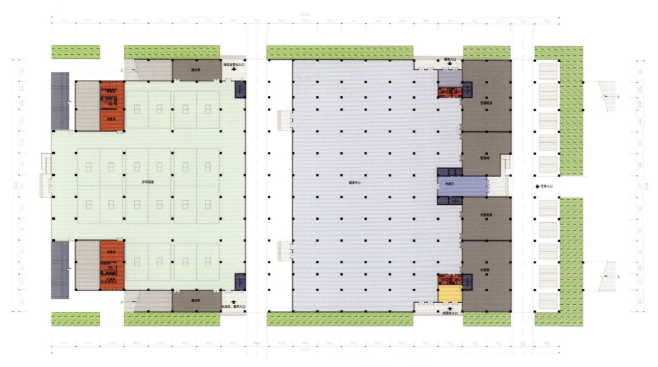

First Floor Plan of Gymnasium 体育馆一层平面图

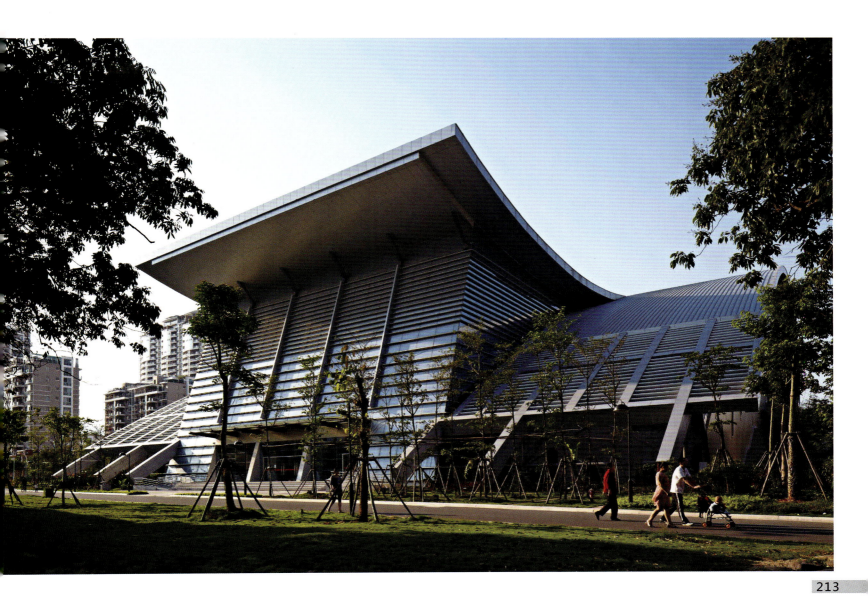

KEY WORDS 关键词

Cantilever Structure 悬挑结构
Creative Elements 创意元素
Ethnic Characteristics 民族特色

Shanxi Sports Center
山西省体育中心

FEATURES 项目亮点

The shape design takes the shape of bass drum, the structure of the lantern and decoration of paper-cut as its creative elements, perfectly reflects the ethnic characteristics, regional features and spirits of the times.

主体育场造型设计以大鼓之形、灯笼之构、剪纸之饰为创意元素，贴切地体现了民族特色、地域特征与时代精神。

Location: Taiyuan, Shanxi, China
Architectural Design: CCDI
Land Area: 730, 340 m²
Total Floor Area: 191, 401 m²
Stadium: 90,066 m²
Gymnasium: 37,791 m²
Natatorium: 29,470 m²
Bicycle Stadium: 17,508 m²
Training Stadium: 16,566 m²
Completion: May, 2011

项目地点：中国山西省太原市
设计单位：中建国际（深圳）设计顾问有限公司
用地面积：730 340 m²
总建筑面积：191 401 m²
体育场：90 066 m²
体育馆：37 791 m²
游泳跳水馆：29 470 m²
自行车馆：17 508 m²
训练馆：16 566 m²
竣工时间：2011年5月

Architectural Design

The designers try to extend the profound culture of Shanxi to architecture vocabulary via modern design techniques, the conception of the main stadium is derived from Shanxi folklore "red lanterns" and "bass drum". The shape design of the building takes the shape of bass drum, the structure of the lantern and decoration of paper-cut as its creative elements, perfectly reflects ethnic characteristics, regional features and spirits of the times. The awning of the stadium employs triangular space-tube-truss cantilever structure, steel structure members and adornment materials become a unified entity, which express concise but powerful characteristics of sports building. The sports center is widely recognized and loved by local people, and it is called as "Red Lantern Stadium" in naming contest among masses.

建筑设计

设计力图将山西深厚的文化通过现代设计手段融入建筑语汇中，主体育场的构思源于对山西民俗特色的"大红灯笼"和"大鼓"的印象。造型设计以大鼓之形、灯笼之构、剪纸之饰为创意元素，贴切地体现了民族特色、地域特征与时代精神。体育场罩棚采用三角形管桁架空间悬挑结构，钢结构构件与外装饰材料浑然一体，体现了简约而有力的体育建筑性格特征。建成后，体育中心得到当地群众的广泛认同和喜爱，在群众征名活动中被称为"红灯笼体育场"。

Site Plan 总平面图

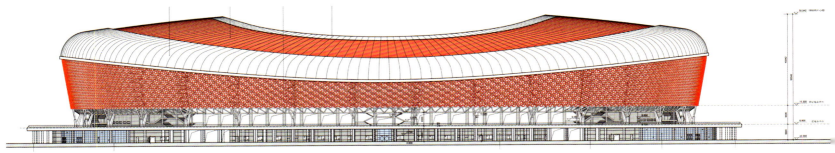
Elevation 1 立面图 1

Elevation 2 立面图 2

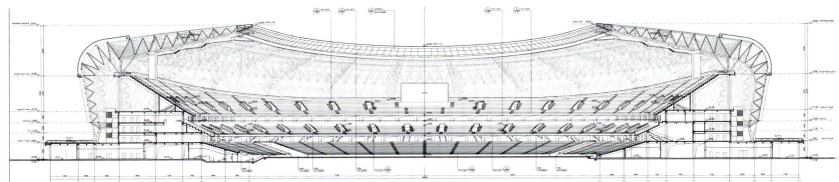
剖面图 1　Section 1

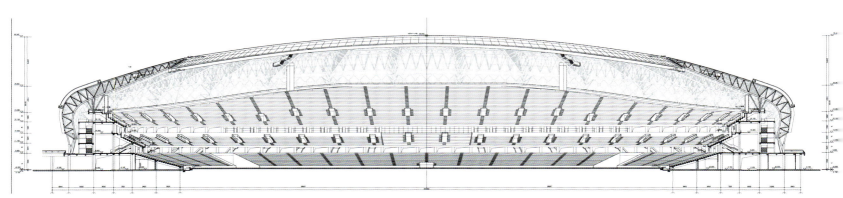
剖面图 2　Section 2

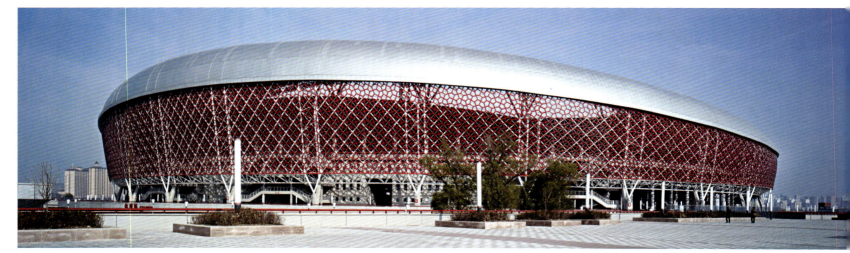

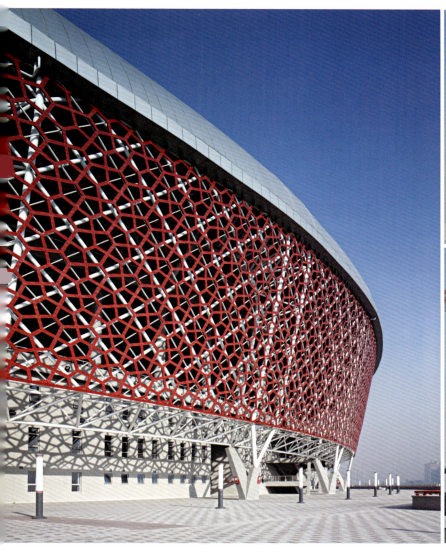
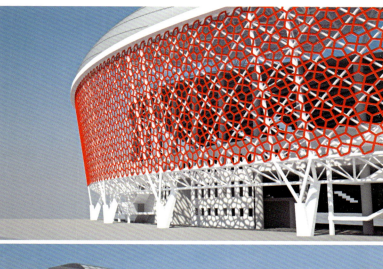
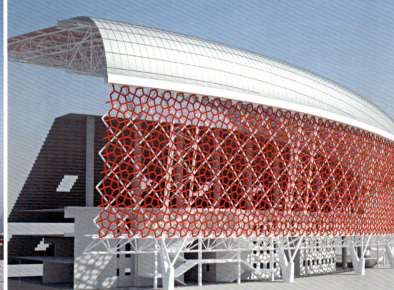
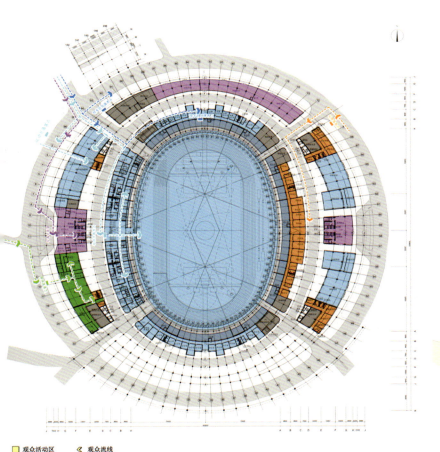
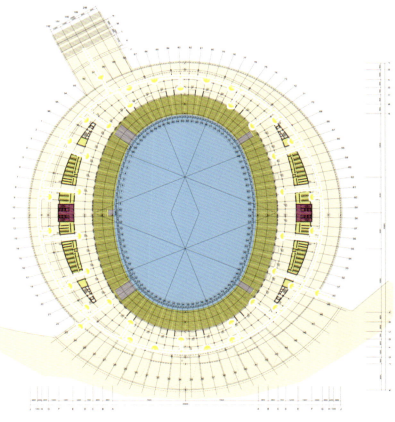

观众活动区	观众流线		
体育竞赛区	运动员流线		
新闻运行区	新闻记者流线		
场馆运行区	场馆运行流线	First Floor Plan 一层平面图	Second Floor Plan 二层平面图
场馆礼宾区	场馆礼宾流线		
电视转播区	电视转播商流线		
安保及交通运行区	安保流线		
比赛场地区			
非赛时使用空间区			

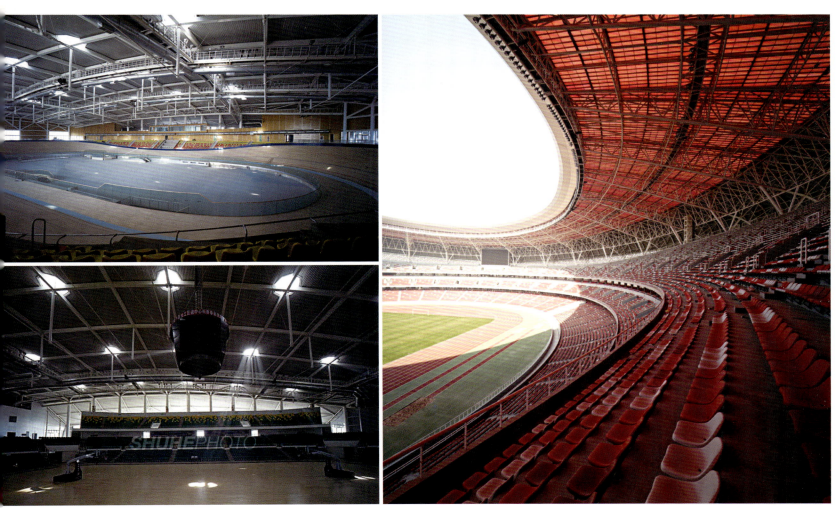

223

KEY WORDS 关键词

Crystal Architectural Appearance 水晶石建筑外观
Luminous Curtain Wall 外观发光幕墙
Poetic Landscape 山水意境

Shenzhen University Games Center Stadium
深圳大运中心体育场

FEATURES 项目亮点

These expressive building blocks and soft landscapes develop into a dialogue, create high recognized overall pattern for university games sports center.

富于表现力的建筑体块与柔和的景观造型之间形成了对话，为大运中心体育场塑造出了具有高度识别性的整体形态。

Location: Aoti Newtown, Longgang District, Shenzhen, Guangdong, China
Architectural Design: Shenzhen Architectural Design & Research Institute Co., Ltd
Land Area: about 36,600 m^2
Floor Area: 135,900 m^2
Building Density: 15.10%
Building Height: 51.30 m
Plot Ratio: 0.35
Green Coverage Ratio: 33.49%
Completion: 2010

项目地点：中国广东省深圳市龙岗区奥体新城
设计单位：深圳建筑设计研究总院有限公司
占地面积：约 36 600 m^2
建筑面积：135 900 m^2
建筑密度：15.10%
建筑高度：51.30 m
容积率：0.35
绿地率：33.49%
竣工时间：2010 年

Overview

Shenzhen University Games Center Stadium is located in Aoti Newtown, Longgang District, Shenzhen, with the land area of 36,600 m^2 and the floor area of 135,900 m^2, it can accommodate 60,000 audiences; it has a six-story sports building, among which four aboveground and two underground

项目概况

深圳大运中心体育场位于深圳市龙岗区奥体新城，占地面积约3.66万 m^2，建筑面积13.59万 m^2，可容纳6万位观众；体育场建筑共6层，其中地下2层，地上4层。

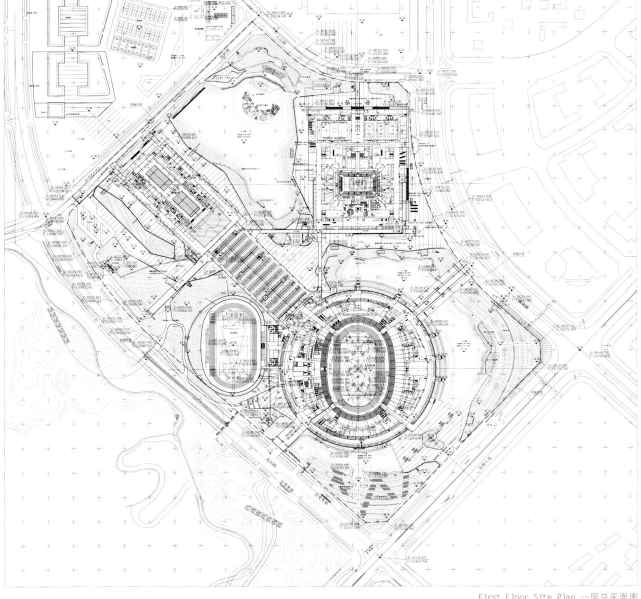

First Floor Site Plan 一层总平面图

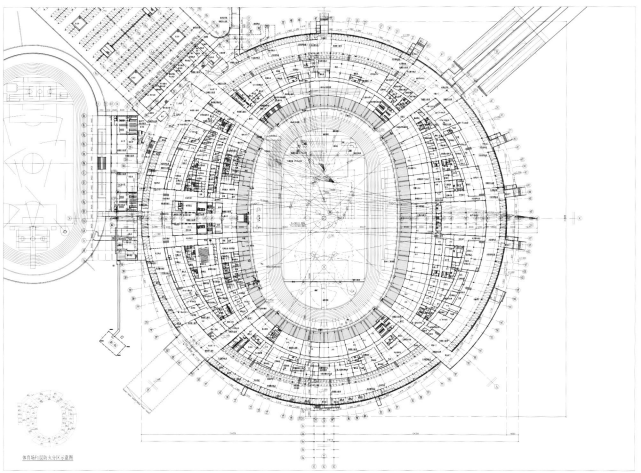

First Floor Plan 一层平面图

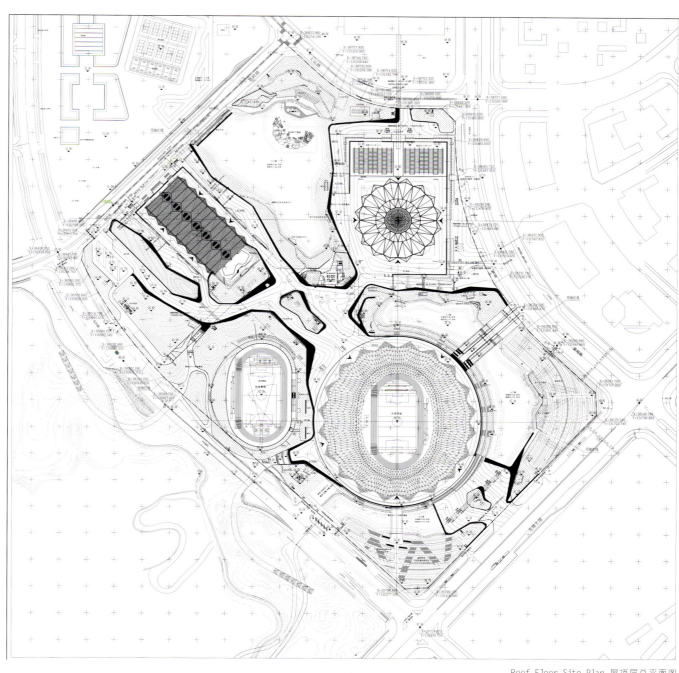

Roof Floor Site Plan 屋顶层总平面图

227

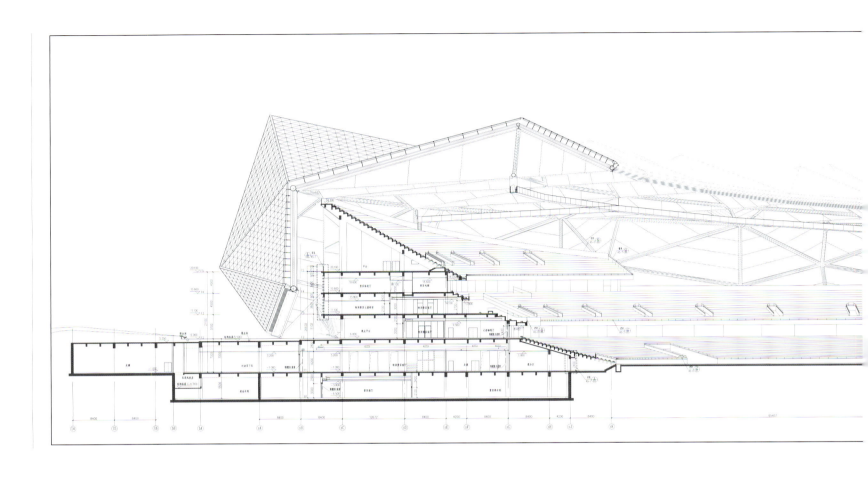

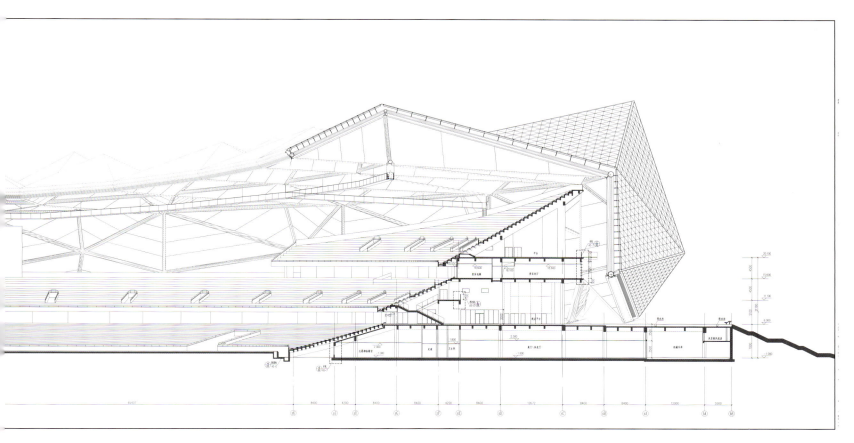

East And West Side Sectional 东西轴剖面图

Architectural Design

The appearance of stadium and gymnasium is crystal stone. A large area landscape park is designed around the building group; this park possesses traditional park's characteristics; the building group exposures to the landscape park, surrounded by soft water. These expressive building blocks and soft landscapes develop into a dialogue, and create high recognized overall pattern for university games sports center. At night, the semi-transparent luminous curtain wall looks like shinning crystal stone, and foils strong festive atmosphere and symbol characteristics; shinning gemstones invert their reflections in the water, which develops into traditional Chinese unique poetic landscape with "mountain", "water" and "stone" as its elements.

建筑设计

体育场馆设计为类似的水晶石建筑外观。这一建筑群周边也设计了面积广阔的景观公园，该公园具备了传统景观园林的特征，建筑群置身于景观园林中，被蜿蜒的水体布环绕着。这些富于表现力的建筑体块与柔和的景观造型之间形成了对话，为大运中心体育场塑造出了具有高度识别性的整体形态。夜间，半透明的发光幕墙，看上去如同熠熠生辉的水晶石，烘托出浓郁的节日氛围和标志特色；湖边，一颗颗发光的宝石倒映在湖面上，波光粼粼，形成传统中国山水画中独特的"山"、"水"、"石"景观意境。

KEY WORDS 关键词

- Curved Shape 曲线造型
- Lingnan Culture 岭南文化
- Cable-membrane Structured Roof 索膜结构屋面

Shenzhen Bao'an Stadium
深圳市宝安体育场

FEATURES 项目亮点

The concept of bamboo forest fully reflects the cultural characteristics in Lingnan district; cable-membrane structured roof expresses the light and ingenious cultural disposition of southern architecture.

体育场中竹林的设计构思充分体现了岭南地域文化特色，索膜结构屋面彰显出轻盈、灵巧的南方建筑文化气质。

Location: Bao'an District, Shenzhen, Guangdong, China
Architectural Design: South China University of Technology Architectural Design and Research Institute
Land Area: 119,700 m²
Total Floor Area: 97,712 m²
Floor Area Aboveground: 71,539 m²
Floor Area Underground: 26,173 m²
Building Height: 55 m
Completion: 2011

项目地点：中国广东省深圳市宝安区
设计单位：华南理工大学建筑设计研究院
用地面积：119 700 m²
总建筑面积：97 712 m²
地上建筑面积：71 539 m²
地下建筑面积：26 173 m²
建筑总高度：55 m
竣工时间：2011 年

Overview

Shenzhen is located in subtropics, gardens can be seen everywhere, the lush bamboos are typical plants in southern China, bamboo forest is the inspiration of building design. The green column that enclose the hall in Bao'an Stadium rightly expresses the irregular light and shadow pattern of bamboo forest, and supports the load of roof and upper grandstand. Spectators enter into the stadium from the space between green columns, and then arrive at lower grandstand, hold a panoramic view of the grand scene in the stadium. The interior design of VIP hall embodies another kind of abstract prospect of bamboo forest.

项目概况

深圳地处亚热带，举步皆为园林，而园林中的葱郁竹林则是中国南方的代表性植物，在项目中竹林演变为建筑设计的灵感。宝安体育场环绕大厅的绿柱正好演绎了竹林不规则的光影图案，并支撑了屋顶和上部看台的荷载。观众从绿柱之间进入体育场，抵达较低处的看台，将整个体育馆的壮观景象尽收眼底。贵宾厅的室内装修则体现了另外一种抽象的竹林意境。

Design Concept

The concept of bamboo forest fully reflects the cultural characteristics in Lingnan district; cable-membrane structured roof expresses the light and ingenious cultural disposition of southern architecture.

设计构思

竹林的设计构思充分体现了岭南地域文化特色，索膜结构屋面彰显出轻盈、灵巧的南方建筑文化气质。

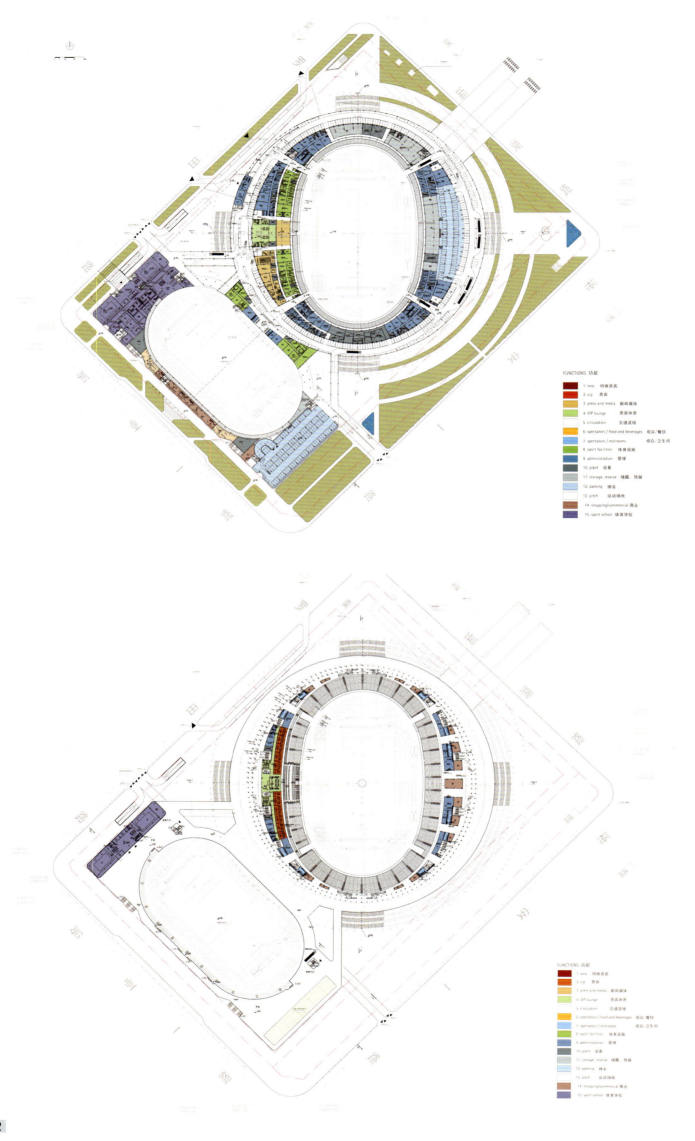

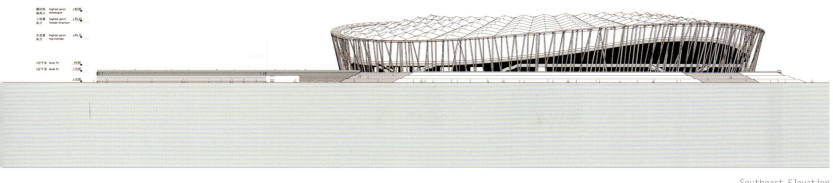

Southeast Elevation
东南立面图

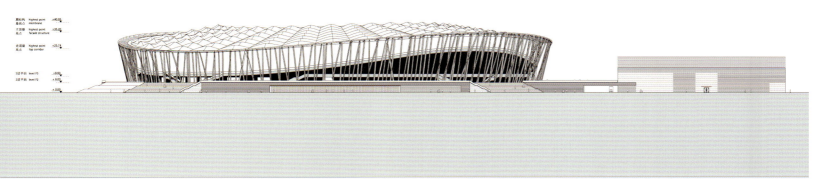

Northwest Elevation
西北立面图

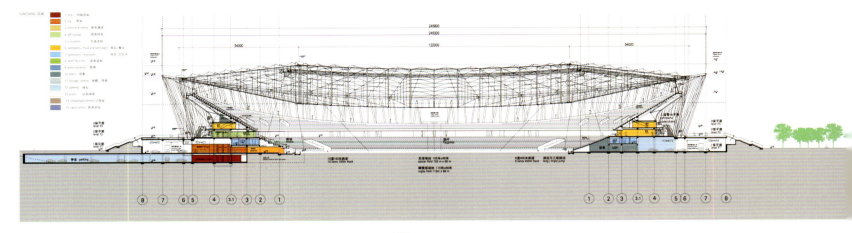

剖面图 1 Section 1

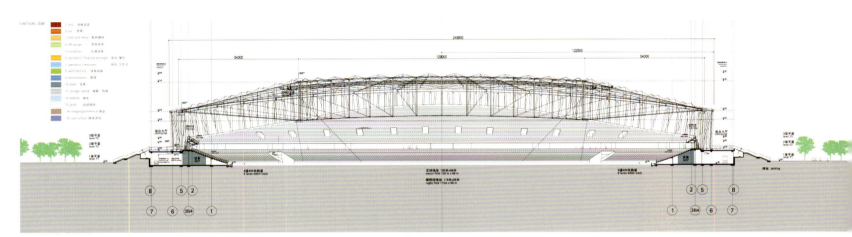

剖面图 2 Section 2

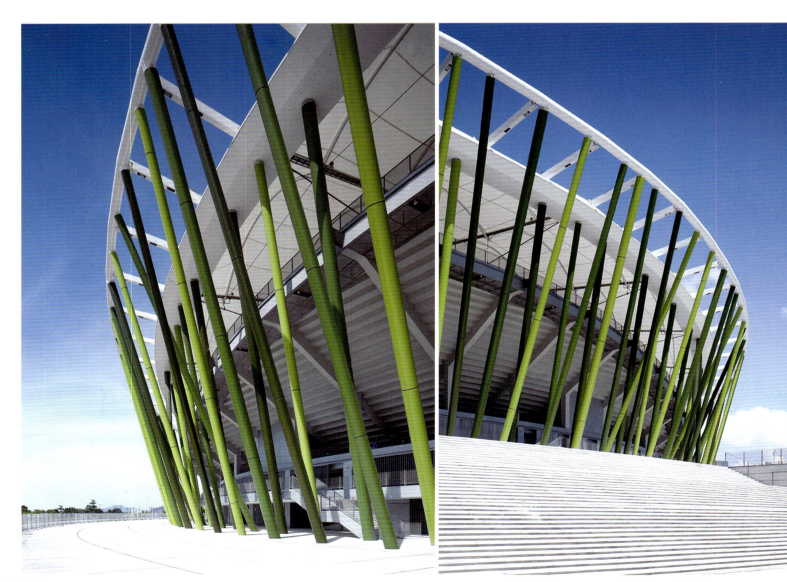

Architectural Design

The roof structure of the stadium keeps the curved shape that upper grandstand is undulate; this kind of shape is lower in south and north, higher in east and west. As wide membrane structure is used in roof coverage and high plasticity of the structure, the whole roof coverage looks like a huge floating cloud hovering on the bamboo forest; hence, unique building appearance is created.

建筑设计

体育场的屋盖构造遵循了上部看台起伏的曲线造型，这一造型在南面和北面较低，而东面和西面则较高。由于体育场屋盖采用宽大的膜结构，以及构造所具备的高度塑性，因此整个屋盖看上去犹如一朵巨大的浮云升腾于竹林之上，由此塑造了独特的建筑外观。

KEY WORDS 关键词	Completed Equipment 功能完备
	Flexible Layout 布局灵活
	Energy Saving 节能环保

Fujian Provincial Physical Rehabilitation and Employment Training Center for the Disabled
福建省残疾人体育康复就业培训中心

FEATURES 项目亮点

This design has improved resource utilization by applying green building materials and relative theories and technologies of construction energy saving, and has also made full use of natural light and natural ventilation to meet the demands of energy saving and to reduce the cost of construction and operation.

设计充分利用绿色环保建材，充分运用建筑节能理论和技术，提高资源利用率，利用天然采光和自然通风，满足节能要求，降低建设和运营成本。

Location: Fuzhou, Fujian, China
Architectural Design: Fujian Provincial Institute of Architectural Design and Research
Land Area: 57,410 m^2
Total Floor Area: 24,448.2 m^2
Floor Area Aboveground: 21,708 m^2
Floor Area Underground: 2,780.2 m^2
Total Floor Area Counted for Plot Ratio: 21,708 m^2
Total Land Area: 6,018.6 m^2
Building Density: 10.48%
Plot Ratio: 0.38
Green Coverage Ratio: 45.5%

Overview

The project is next to Zhuangyuanzhi Road, at Jianxin Town, Cangshan District in Fuzhou City. Its north is close to Hua Hao Yue Yuan Community and its east faces with Fujian Provincial Armed Police Command School in Fuzhou; meanwhile, its south is mountain and its west is adjacent with vegetable patches.

项目地点：中国福建省福州市
设计单位：福建省建筑设计研究院
用地面积：57 410.0 m^2
总建筑面积：24 448.2 m^2
地上建筑面积：21 708 m^2
地下建筑面积：2 780.2 m^2
计容建筑总面积：21 708 m^2
总占地面积：6 018.6 m^2
建筑密度：10.48%
容积率：0.38
绿地率：45.5%

项目概况

项目位于福州市仓山区建新镇，紧邻状元支路，地块北邻花好月圆小区，东面与省武警福州指挥学校相望，南面为山地，西临菜地。

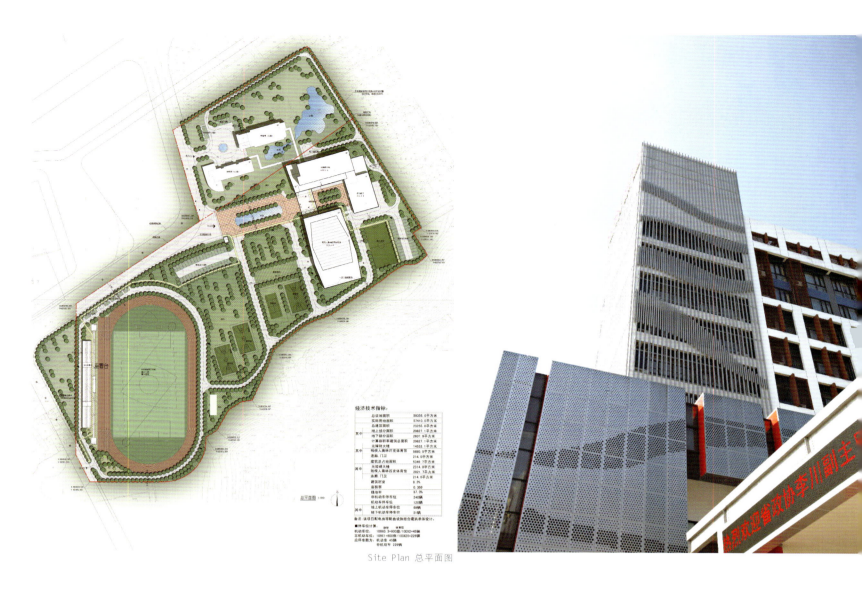

Site Plan 总平面图

Architectural Design

This center is a good place for the disabled to live, practice and rehabilitate, which strengthens the link among greening, landscape and the building. This design has improved resource utilization by applying green building materials and relative theories and technologies of construction energy saving, and has also made full use of natural light and natural ventilation to meet the demands of energy saving and to reduce the cost of construction and operation. Fully functional – completed sports facilities to meet demands of international or domestic individual sport games. Comprehensive effect Reflected – flexible layout fits in with requirements of the use of various functions, together with the operational needs, so as to enhance the center its own operation capacity. The combination of long-term and short-term construction projects and the scientific use of land resources have fully reflected Comprehensive, coordinated and sustainable development of the concept of scientific development, so as to leave room for possible reconstruction and development. The reasonable arrangement for human, traffic and information makes them work independently and non-interfering.

建筑设计

本中心是残疾人生活、运动、康复的乐园，强化绿化、景观与建筑的紧密结合。设计充分利用绿色环保建材，充分运用建筑节能理论和技术，提高资源利用率，利用天然采光和自然通风，满足节能要求，降低建设和运营成本。功能基本完备——齐全的竞赛设施，满足国际国内单项体育比赛的要求。体现综合效应——布局灵活，可适应多种功能的使用需求，具有可经营性，坚持"以馆养馆"以增强体育馆的自身经营能力。结合远、近期建设项目，科学利用土地资源，充分体现全面、协调、可持续发展的科学发展观，为可能的改建和发展留有余地。科学地安排人流、车流、信息流，做到"道路宽畅、人车分道、管网下地、网线分道、互不干扰"。

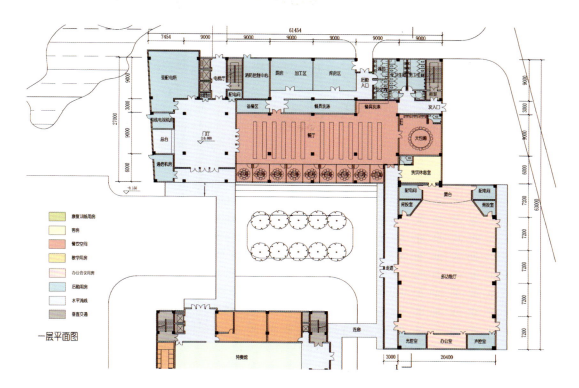

First Floor Plan 一层平面图

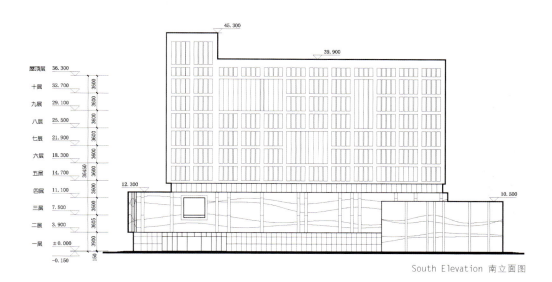

South Elevation 南立面图

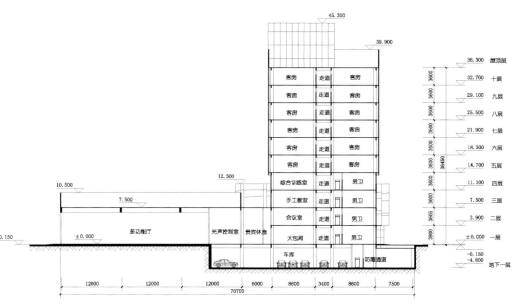

Sectional Drawing 剖面图

241

KEY WORDS 关键词

- Steel Structure 钢结构
- Integrated Platform System 一体化平台系统
- Canopy 罩棚设计

Dalian Stadium
大连市体育场

FEATURES 项目亮点

The canopy and stand correspond to each other in height, form and function. ETFE inflatable pillows are used on the cover of the canopy, which acts as a giant illuminant sparkling in different colors in the night, creating a strong atmosphere of an urban landscape.

体育场罩棚造型与看台高度趋于契合，形式与功能完整、统一，罩棚表面采用 ETFE 充气枕形式，通过气枕色彩及质感的变化，营造出浓郁的城市景观氛围。

Location: Dalian, Liaoning, China
Architectural Design: The Architectural Design & Research Institute of HIT
Base Area: 80,650 m²
Total Floor Area: 119,622 m²

项目地点：中国辽宁省大连市
设计单位：哈尔滨工业大学建筑设计院
建筑基底面积：80 650 m²
总建筑面积：119 622 m²

Overview

Located at Ganjingzi District in Dalian, Dalian Sports Center is on the east of Xibei Road and on the north of Lanling Road, covering an area of 820,000 m². As the core building in this center, Dalian Stadium can accommodate 60,000 people with a total floor area of 120,000 m², and is known as a superfine sports building.

项目概况

大连体育中心用地位于大连市甘井子区，西临西北路，南临岚岭路，占地 82 万 m²。体育场为体育中心核心建筑。体育场能够容纳 6 万观众，总建筑面积 12 万 m²，为特级体育建筑。

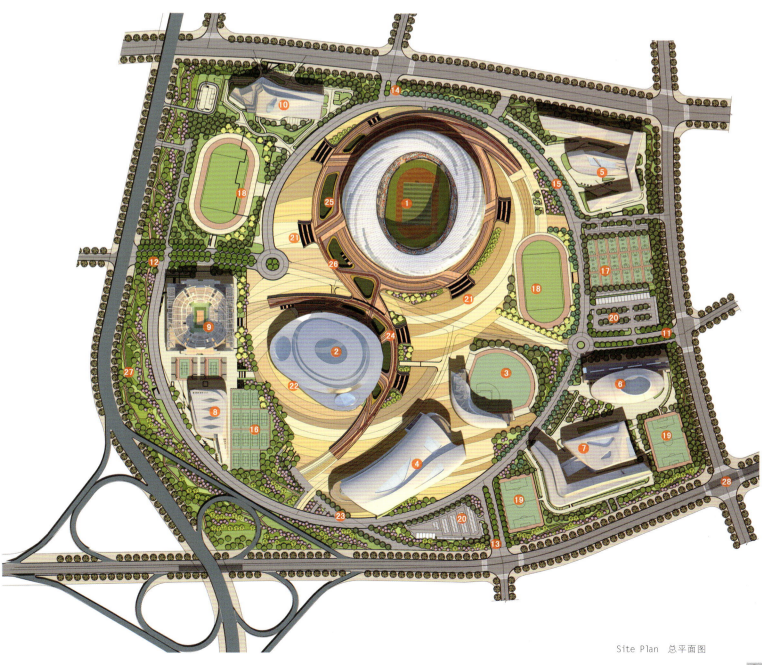

Site Plan 总平面图

244

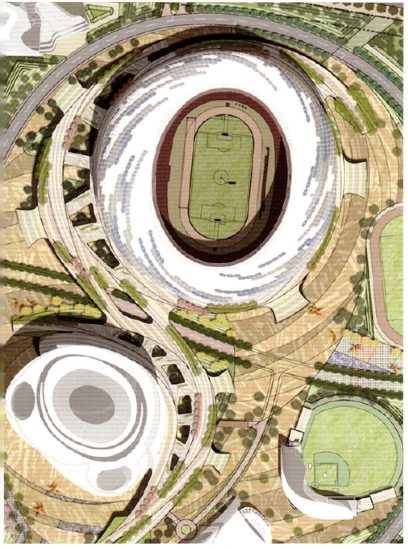

Design Concept

Designers worked in concert with the urban culture of Dalian; took blue and white as the basic colors, which symbolize ocean and interstellar space nurturing life endlessly. The canopy and stand correspond to each other in height, form and function. ETFE inflatable pillows are used on the cover of the canopy, which acts as a giant illuminant sparkling in different colors in the night, creating a strong atmosphere of an urban landscape.

设计理念

体育场设计理念力图与大连的城市文化相呼应，以蓝白色为基本色调，象征海洋与星际，孕育生命、循环不息、运动不止。体育场罩棚造型与看台高度趋于契合，形式与功能完整、统一，罩棚表面采用 ETFE 充气枕形式，通过气枕色彩及质感的变化，形成与体育馆异曲同工的动感态势，夜晚变幻的气枕罩棚如同巨型的发光体，营造出浓郁的城市景观氛围。

Architectural Design

The platform that connects the stadium and the gymnasium runs through the entire sports center from south to north, forming an aerial corridor which breaks the single traffic function of the platform for traditional sports building and presents a sports park of three-dimensional landscape. A part of the platform is filled with soil to grow plants and transparency is situated in proper positions to interact with the plants on the ground. Tracks stagger on the platform, which forms a jogging trail that surrounds the stadium. The integrated platform system weakened the construction scale of the stadium and the gymnasium, making the large-span roof of the building link up with the ground closely. It is a real public urban sport space where people can stroll and stay freely.

The canopy occupies an area of 68,000 m², and is composed of 2,736 inflatable pillows. The grooves between the pillows serve as gutters that discharge the rainwater down along the main steel structure. PEFE grid film is disposed inside the pillow for sunshade, and the light transmittance is 30%, which transforms the light into natural diffused light to prevent the audience from dazzling and covers the complex steel structure, various pipes and so on, so as to create an integrated visual impact for the interior space.

建筑设计

体育场与体育馆的连接平台贯穿整个体育中心南北，形成空中交通走廊，设计突破传统体育建筑连接平台的单一的交通功能，将整个平台系统设计成空中立体景观运动公园。平台上部局部采用填土，形成人工绿丘，种植绿化，并在恰当的位置设置透空，将地面层植被渗透到平台标高的空中。平台面层以跑道形式交错分隔，形成环绕场馆的缓跑径。一体化的设计弱化了体育馆及体育场的建筑尺度，使得大跨度的建筑屋盖好似与大地亲密衔接，自然生成，市民可以漫步畅游，亦可停留休憩，真正创造了属于公众的城市运动空间。

体育场罩棚面积 68 000 m²，共用 2 736 块气枕，气枕间的结构凹槽作为排水沟，将雨水顺着主体钢结构自然排出。气枕内侧设置 PEFE 网格膜进行遮阳，透光率为 30%，光线入射后成为自然漫射光，防止观众炫目，并将复杂的钢结构、各种管道等进行遮挡，使内场形成完整的空间感观效果。

KEY WORDS 关键词

Unique Shape 造型独特

Large-scale Structure 大跨度结构

Times Flavors 时代气息

Fujian Physical Polytechnic College Track and Field Training Center
福建省体育职业技术学院田径训练馆

FEATURES 项目亮点

The main body used technique of contrast, creating a integrated free space from roof to the wall by utilizing roofing curve. Facade modeling based on functional perspective and through wall division and application of emptiness the reality to develop into a strong visual effect.

建筑主体采用对比的手法，运用屋面曲线，塑造从屋面到墙体一体化的自由空间。立面造型从功能角度出发，通过对墙面的划分和虚实手法的应用，形成强烈的视觉效果

Location: Fuzhou, Fujian, China
Architectural Design: Fujian Architectural Design and Research Institute
Land Area: 28,090 m²
Total Floor Area: 16,198.8 m²
Building Density: 34.1%
Plot Ratio: 0.57
Green Coverage Ratio: 30.1%

项目地点：中国福建省福州市
设计单位：福建省建筑设计研究院
用地面积：28 090 m²
总建筑面积：16 198.8 m²
建筑密度：34.1%
容积率：0.57
绿地率：30.1%

Overview

Fujian Physical Polytechnic College is located in Fuzhou, and within Fujian provincial sports center. The existing tennis court is in the southern side, the swimming pool and outdoor ball games facilities is in the eastern side, the western side echoes to the existing three comprehensive training buildings across the planned road. The main part of this tranning hall is large space of track and field training venue, which is used as daily indoor training room for the track team.

项目概况

福建省体育职业技术学院田径训练馆，位于福州市福建省体育中心总用地之内。南侧为现有网球馆，东侧为游泳馆及室外球类运动设施，西侧隔着规划路与已建三幢综合训练馆相呼应。场馆主体部分为大空间田径训练场，为田径队日常室内训练使用。

Architectural Design

The shape of this building echoes to the surrounding large sports building, and strive to create a whole sports community space. The structural rationality of functional space is highlighted, emphasizing on smooth and stretched architecture form. The main body used technique of contrast, creating a integrated free space from roof to the wall by utilizing roofing curve. Facade modeling based on functional perspective and through wall division and application of emptiness the reality to develop into a strong visual effect. Horizontal shading lattice combined with building body perfectly, which can prevent glare of sunlight on the one hand, and on the other hand give times flavors to the building benefiting from its plump shape and light line.

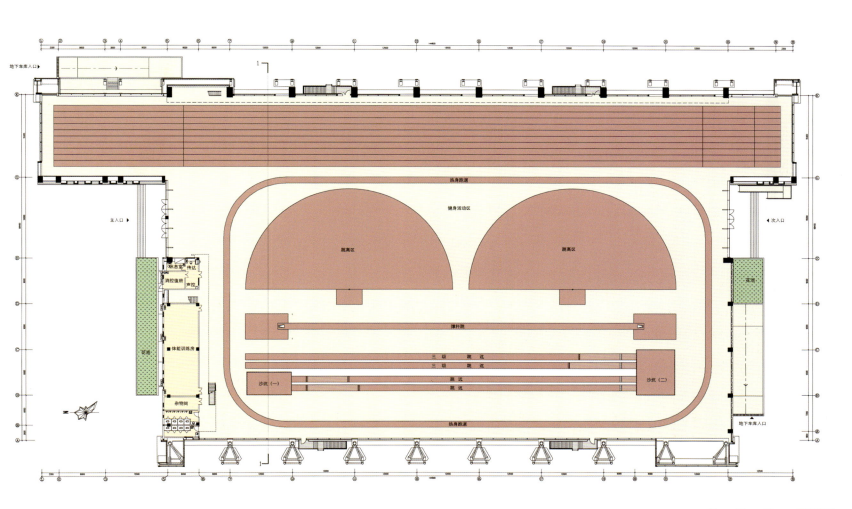

First Floor Plan 一层平面图

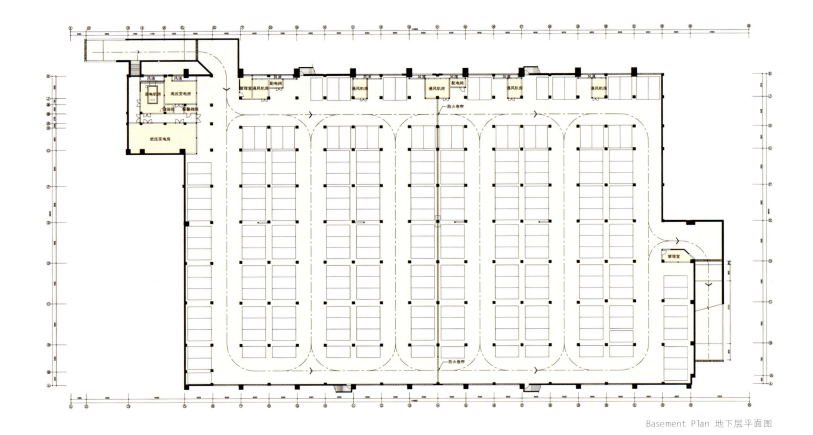

Basement Plan 地下层平面图

建筑设计

建筑形体与周边体育中心大体量建筑相呼应，力求创造出整体化的体育群落空间，突出功能空间的结构理性，着重强化流畅、舒展的建筑形态。建筑主体采用对比的手法，运用屋面曲线，塑造从屋面到墙体一体化的自由空间。立面造型从功能角度出发，通过对墙面的划分和虚实手法的应用，形成强烈的视觉效果。水平遮阳格片与建筑型体完美组合，一方面防止日照眩光，另一方面以其饱满的形体和轻灵的线条，赋予建筑强烈的时代气息。

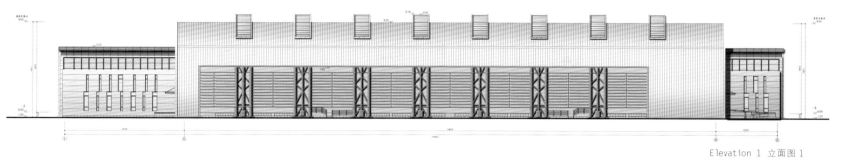

Elevation 1 立面图 1

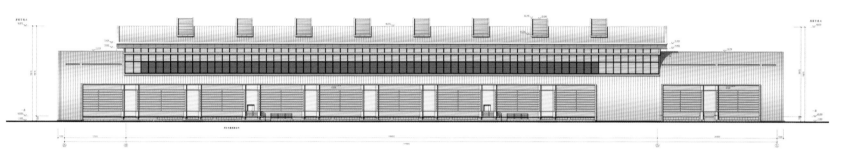

Elevation 2 立面图 2

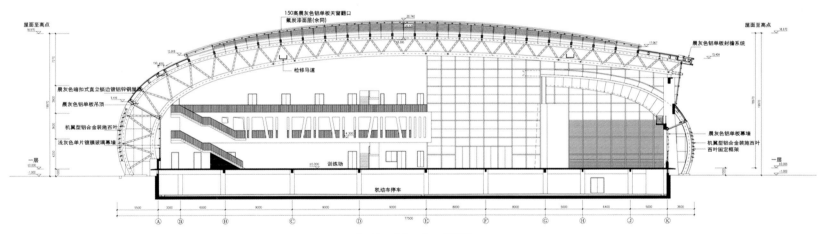

Sectional Drawing 剖面图

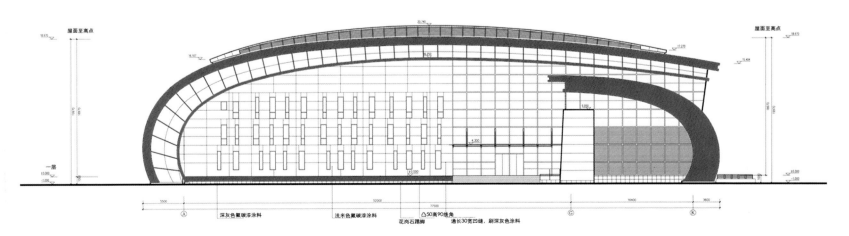

Elevation 立面图

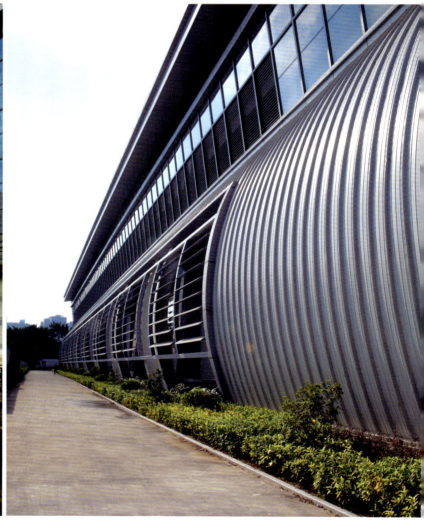

KEY WORDS 关键词

- **Practical Economy** 实用经济
- **Moderate Innovation** 适度创新
- **Green Building** 绿色建筑

Liaoning Sports Training Center (Baiye Base)
辽宁省体育训练中心（柏叶基地）

FEATURES 项目亮点

The project design follows the local conditions, realistic, practical, economical and moderately innovative principles, by using of new materials and new technology to achieve green building and land materials, energy and water conservation goals.

项目遵循因地制宜、实事求是、实用经济、适度创新的设计原则，采用新材料新技术实现绿色建筑节地、节材、节能、节水的目标。

Location: Shenyang, Liaoning, China
Architectural Design: Liaoning Provincial Institute of Architectural Design
Land Area: about 674,900 m²

Overview

The main goal of this project is to meet the Liaoning Provincial Sports Bureau's shooting, archery, cycling and swimming, fencing, football training, and the 12th National Games shooting, archery, cycling competitions and needed supporting facilities in 2013. Creating an ecological landscape of international sports training base is the design purpose.

项目地点：中国辽宁省沈阳市
设计单位：辽宁省建筑设计研究院
用地面积：约 67.49 万 m²

项目概况

本工程主要功能为满足辽宁省体育局射击、射箭、自行车以及游泳、击剑、足球等训练和2013年第12届全运会射击、射箭、自行车比赛以及比赛配套的使用需求。打造生态化的国际化体育训练基地，是本次设计的宗旨。

综合馆

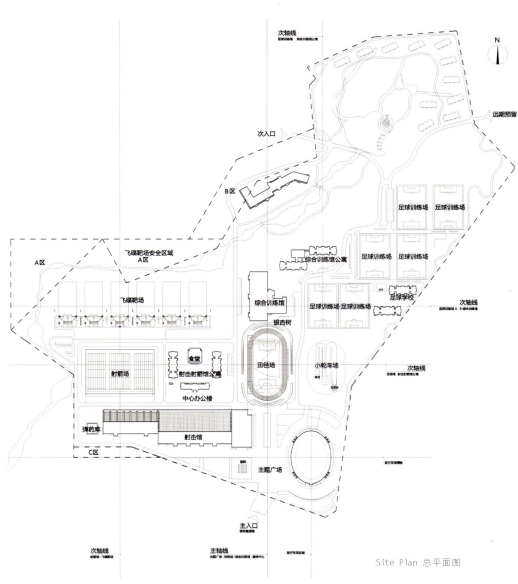

Site Plan 总平面图

Green Building

The project design follows the local conditions, realistic, practical, economical and moderately innovative principles, by the use of new materials and new technology to achieve green building and land, materials, energy and water conservation goals.

The project uses a number of advanced energy-saving technologies, building envelope energy-saving technologies; utilize hot design principles of natural ventilation systems, solar panels, breathing walls, phase change thermal storage materials; solar water heating technology, solar power technology; intelligent control technology; rainwater harvesting techniques; roof drainage system to collect rainwater flow pressure, drainage layer technology, water treatment, water reuse technology; construction of a new heat recovery ventilation system technology, energy efficient lighting; water-saving sanitary equipment; energy-saving air conditioning technology.

绿色建筑

项目遵循因地制宜、实事求是、实用经济、适度创新的设计原则，采用新材料新技术实现绿色建筑节地、节材、节能、节水的目标。项目采用多项先进的节能技术，建筑围护结构节能技术；利用热压原理设计自然通风系统、太阳能集热板、呼吸式幕墙、相变蓄热材料；太阳能热水技术、太阳能发电技术；智能控制技术；雨水收集利用技术；屋面雨水压力流收集排水系统；同层排水技术；水处理、中水回用技术；建筑新风系统技术热回收装置；高效节能灯具；节水型卫生设备；节能空调技术。

自行车馆